Lecture Notes in Computer Science 1260

Edited by G. Goos, J. Hartmanis and J. van Leeuwen

Advisory Board: W. Brauer   D. Gries   J. Stoer

Springer
Berlin
Heidelberg
New York
Barcelona
Budapest
Hong Kong
London
Milan
Paris
Santa Clara
Singapore
Tokyo

Darrell Raymond   Derick Wood
Sheng Yu   (Eds.)

Automata Implementation

First International Workshop
on Implementing Automata, WIA'96
London, Ontario, Canada, August 29-31, 1996
Revised Papers

 Springer

Series Editors

Gerhard Goos, Karlsruhe University, Germany

Juris Hartmanis, Cornell University, NY, USA

Jan van Leeuwen, Utrecht University, The Netherlands

Volume Editors

Darrell Raymond
Sheng Yu
University of Western Ontario, Department of Computer Science
London, Ontario, Canada N6A 5B7
E-mail: drraymon@porsche.csd.uwo.ca
        syu@csd.uwo.ca

Derick Wood
The Hong Kong University of Science and Technology
Department of Computer Science
Clear Water Bay, Kowloon, Hong Kong
E-mail: dwood@cs.ust.hk

Cataloging-in-Publication data applied for

Die Deutsche Bibliothek - CIP-Einheitsaufnahme

Automata implementation : revised papers / First International
Workshop on Implementing Automata, WIA '96, London, Ontario,
Canada, August 29 - 31, 1996. Darrell Raymond ... (ed.). - Berlin ;
Heidelberg ; New York ; Barcelona ; Budapest ; Hong Kong ;
London ; Milan ; Paris ; Santa Clara ; Singapore ; Tokyo : Springer,
1997
  (Lecture notes in computer science ; Vol. 1260)
  ISBN 3-540-63174-7

CR Subject Classification (1991): F.4.3, F.3, F.1.1

ISSN 0302-9743
ISBN 3-540-63174-7 Springer-Verlag Berlin Heidelberg New York

© Springer-Verlag Berlin Heidelberg 1997
Printed in Germany

Typesetting: Camera-ready by author
SPIN 10549844   06/3142 – 5 4 3 2 1 0   Printed on acid-free paper

Foreword

The papers contained in this volume were presented at the first international Workshop on Implementing Automata, held August 29–31, 1996, at the University of Western Ontario, London, Ontario, Canada.

The Workshop addressed issues involved in the implementation of automata of all types. The goal of the Workshops is to provide a forum for active researchers who are involved in various aspects of this area. Clearly, the area overlaps other areas such as the natural-language processing, pattern matching, speaker recognition, and VLSI testing, to name a few.

The major motivation in starting these workshops is that there has been no single forum in which automata-implementation issues are discussed. The interest shown in the first workshop demonstrates that there is a need for such a forum. Two recent reports by members of the Theoretical Computer Science community suggested that researchers should engage in more application-oriented research. The number of packages, toolkits, and systems that have recently been developed to manipulate formal-language-theory objects can be interpreted as a response to this call. Regular expressions are used in many utilities and systems; indeed, they are such a useful tool that variants of regular expressions have been introduced in document processing, in concurrent programming, and in database systems. Finite-state systems are used to specify protocols, to model faults in VLSI circuits, and to solve spaghetti-programming problems. Finite-state transducers are a cornerstone of work on phonology and morphology.

The general organization and orientation of WIA conferences are governed by a Steering Committee composed of Stuart Margolis, Denis Maurel, Sheng Yu, and Derick Wood. The 1997 meeting of WIA will also take place at the University of Western Ontario and the 1998 meeting will be held in France.

The local arrangements for WIA 96 were co-chaired by Darrell R. Raymond and Sheng Yu who were ably assisted by Caroline Sori, the conference coordinator. The conference was supported in part by the Natural Sciences and Engineering Research Council of Canada and by the University of Western Ontario. We gratefully acknowledge the efforts of all members of the local arrangements team and last, but certainly not least, we thank the researchers who were brave enough to demonstrate their software.

The Program Committee for WIA 96 was:

Franz Guenthner	Ludwig Maximillian Universität, Germany
Boleslaw Mikolajczak	University of Massachusetts, USA
Darrell Raymond (Co-Chair)	University of Western Ontario, Canada
Susan Rodger	Duke University, USA
Kai Salomaa	University of Turku, Finland
Jorge Stolfi	Universidade de Campinas, Brazil
Bruce Watson	Eindhoven University, the Netherlands
Derick Wood	HK Univ. of Science & Technology, Hong Kong
Sheng Yu (Co-Chair)	University of Western Ontario, Canada

Contents

Contents

WIA and the Practice of Theory in Computer Science

Darrell Raymond*

Department of Computer Science
University of Western Ontario
London, Ontario, Canada
N6A 5B7
drraymon@porsche.csd.uwo.ca

This is a revised and expanded version of the opening talk given at WIA '96.

Good morning. Let me welcome you to WIA '96, the first international workshop on implementing automata. This is the first of what I hope will be a continuing series of workshops investigating software for formal language theory. I would like to take a few minutes to explain the origins of this workshop, and to tell you some of my ideas about its philosophy.

Whence WIA? Why the need for a workshop of this type? As there are already many (perhaps too many) computer science conferences and workshops, any new meeting faces a rather stiff need to justify its existence. WIA came about primarily because there is no other good forum for systems that support symbolic computation with automata. Systems that we know about include Howard Johnson's INR, Bruce Watson's Fire Lite, Susan Rodger's FLAP and JFLAP, Champarnaud's AUTOMATE (which has spun off several related projects), AMORE, several different systems called AUTOMATA, Gertjan van Noord's FSA Utilities, our own Grail project, Kartunnen's work at Xerox, Guenthner's work at CIS in Munich, and many others. In addition to the systems I have named, there is a vast amount of applied work, most of it undocumented, using automata for practical applications such as protocol analysis, IC design and testing, telephony, and other situations where automata software is useful.

This is good and interesting work, and it needs a place to be exhibited and discussed. Existing journals and conferences, however, seem to have a difficult time in finding a place for what we do. Theoretical arenas sometimes treat this work as 'mere' implementation, a simple working-out of the algorithms, theorems, and proofs that are the 'real' contribution to the field. Systems-oriented venues, on the other hand, sometimes find this kind of work suspect because it appears to be aimed at theoreticians. It is tricky navigating between the Scylla of the too-abstract and the Charybdis of the too-practical.

Providing a forum for this work is a useful goal, and a sufficient one for WIA. But I think WIA is part of something more fundamental, and a process I want to encourage: the re-appraisal of the value of programming in computer science. Let me briefly describe what I have in mind.

---

* Presently a consultant with The Gateway Group, Waterloo, Ontario.

About five years ago I attended a Dagstuhl workshop entitled 'Implementing Algorithms', organized by Kurt Melhorn and Stefan Nahrer of the Max Planck Institute. The title suggests a practical bent, but in case anyone was confused, Melhorn's opening talk left no room for doubt. Melhorn expressed a concern about the increasingly sterile trend of theoretical work in computational geometry. Too much of it, he said, produces algorithms that have nice analytic properties but are impractical, either because of the complexity of the algorithms or because of their use of dubious assumptions, such as infinite precision arithmetic. Melhorn's goal was to reorient this area, and others, towards practical algorithm design.

Melhorn's observation led me to ponder how the theoretical community ended up pursuing work that was, by any reasonable definition of the word, impractical. Was it always this way? The answer seems to be no. Jurg Nievergelt, another attendee at that Dagstuhl workshop, pointed out that if you look at an old copy of CACM, you find that the articles contain not just algorithms, but source code for programs: a blend of theory and practice used to be the hallmark of computer science. In the not-too-distant past, computer scientists were motivated by practical problems, they created algorithms to address those problems, and they viewed analysis and implementation as being of equal importance. Implementation was essential, because the main motivation of the work was to write a program to solve a practical problem, and not just to publish a result. Many of our best computer scientists show an equal facility in implementation and theory: Knuth and Turing are perhaps the most prominent examples.

Knuth is and Turing was an accomplished generalist, but that kind of spirit appears to be lacking in modern computer science. The trend is rather towards specialization, with a prominent division between practitioners and theoreticians. The latter work on abstract problems with interesting mathematical properties, while the former work on improving actual software systems. This division of labour has been accompanied by the development of serious prejudices: some theoreticians view implementation as of secondary importance, and some practitioners act as if theory were of little relevance to their work.

Why did this division of labour happen? Several reasonable explanations suggest themselves. For one, some people gravitate more naturally towards mathematics and others towards system work. Physics has its experimental and theoretical sides; similarly, one might expect to see both 'experimental' and theoretical computer science. As any discipline grows, it becomes very hard for any one person to keep on top of both practical and theoretical developments. Finally, not everyone in the field thinks that a distinctive characteristic of computer science is its merger of practical and theoretical work, as seems to me evident in the very name of the discipline.

The division of labour does not, by itself, mean that theoretical work was destined to become impractical. Impracticality arose when theoreticians adopted abstract notions that (not always, but too often) didn't adequately model the properties of actual computers dealing with typical problems. The very important theoretical notions of asymptotic complexity and worst-case analysis, for example, can sometimes be weak reeds to lean on in choosing between algorithms. While we can learn a lot about some algorithms by considering their worst-case performance, it is equally true that for other algorithms, worst-case performance bears no resemblance to the typical performance, or even the performance over the vast majority of inputs. A fine example of this is found in Bruce Watson's empirical work on minimization of automata. Brzozowski's algorithm for minimization takes exponential time in the worst case, since it involves using determinization (twice), and thus it appears to be much worse than Hopcroft's algorithm, which takes $O(n \log n)$ time in the worst case. The disparity in worst-case performance is so great that one is tempted to conclude that Brzozowski's algorithm could never be competitive. Yet Watson showed that Brzozowski's algorithm is in practice better than Hopcroft's. Empirical work by Ted Leslie (with Derick Wood and me) suggests that the reason for this is the relatively small number of cases in which determinization produces a result of exponential size.

The notions of asymptotic complexity and worst-case analysis are elegantly employed in the theory of NP-completeness, truly one of the jewels of theoretical computer science. The value of this theory is unquestioned, and I hope no one considers it a challenge if I point out that one of the unfortunate side effects of the theory is to encourage theoreticians to look for results that are of relatively minor interest to practitioners. Mathematicians have long solved problems by reducing them to previously-solved problems; but with the theory of NP-completeness, you could also get a publishable result (and not solve a problem) by reducing a problem to a previously-*un*solved one (a technique also practiced in proofs of undecidability and incomputability). This remarkable idea marked a fundamental divergence in the value structures of theoreticians and practitioners. Crudely speaking, it didn't matter to theoreticians that scheduling needed to be done, because they could show that doing it optimally was intractable. Meanwhile, for practitioners, it didn't matter that optimal scheduling was in general intractable, since they were only interested in reasonable scheduling anyway. For a time, it seemed more popular for theoreticians to publish work showing the intractability of problems, than to work on tractable problems, or tractable variants of intractable problems.

I hope it is understood that my goal is not to denigrate asymptotic complexity, worst-case analysis, or the theory of NP-completeness. These are valuable and powerful ideas, and should continue to be used—in the right hands, and with the right provisos. I merely point out that, when separated from practical work, these tools sometimes lead us to value theorems and proofs that are not of much practical relevance; moreover, the results we gain with these techniques can be misleading.

All this, you may argue, simply means that it's a good idea to buttress our theoretical results with good simulations, and perhaps do some more careful explanation of how to use theoretical notions. For me this is still not the main point, because I believe that programming and experimentation are not means to better theory, but ends in their own right, of comparable status and utility.

For the other aspect of the divide between practitioners and theoreticians is that programming has somehow become a second-class citizen in its own discipline. It seems to me that theoreticians can't discuss programming without moulding it in their own terms, terms which are to me inappropriate.

Consider the celebrated dispute about program verification, touched off by DeMillo, Lipton and Perlis's paper *Social Processes and Proofs of Theorems and Programs*. The bitter dialogue begun by this paper was, in part, an argument about whether programs were more like theorems or proofs. Many mathematicians and philosophers found time to discuss this issue, but to me it was striking that very few real, working programmers were consulted on the topic! To have mathematicians and philosophers describing programming is something like having choreographers describe musicianship; there is a passing similarity between the skills, but we don't really expect ballerinas to understand what it is to play the oboe.

If a programmer *had* been consulted about what constitutes programming, I am willing to bet that the words 'theorem' and 'proof' would be conspicuous by their absence. Programming as it is practised today rarely involves proving anything, and theorems are among the more unused objects in the programmer's toolkit. One might argue that programmers *ought* to make more use of theorems, or that the lack of proof techniques in coding is an abominable state of affairs. I would counter that this is about the same as condemning musicians for not doing stretching exercises. The purported value of mathematical techniques for programming is as yet an unverified claim.

If programs were theorems or proofs, then mathematicians might be the best programmers. I have not conducted a formal study, but I believe that it is easily observed that the ability to transfer skills between mathematics and programming is about as slim as the ability to transfer skills between mathematics and playing music. Naturally, there are mathematicians who are also good programmers, and who also are good musicians. For any skill X, there is probably a non-empty set of mathematicians who are also good at X, but correlation is not causation.

If programming isn't the process of generating theorems or proving them, what is it? In my view, it is one of the few unique contributions of computer science to human knowledge: it is the development and use of formalisms comparable to mathematics in precision and rigor, but with the additional goal of developing a useful artifact. Programming is *not* mathematics, but it is *like* mathematics in certain ways (while in other ways, it is more like experiments in the natural sciences). Some mathematicians will squawk with dismay at my claim of rigor and precision, given the wealth of buggy software and the undisciplined hacker culture that produces so much of it.[2] Notice that I did not say 'correctness', only rigor and precision. Programs may not be correct,[3] but they

---

[2] This same culture also produces some of the finest examples of good code to be found. No one has yet seen fit to draw any claims about causation in this direction, however.

[3] In most cases they cannot be said to be correct or incorrect, since there is typically no specification against which to test for correctness.

are rigorous and precise in the sense that they are unambiguous, objective descriptions of a repeatable process. Mathematicians rightly pride themselves on their careful definitions and proof techniques, but programmers have an advantage in their machines, which automatically check and enforce a high level of rigor. Moreover, the programmer's product is *executable*. Executability is the essential core of programming, for computers exist primarily to do something, not to prove something. Mathematicians accrete proofs; programmers accrete programs. It is only as mathematicians that we are interested in reducing problems to previously-unsolved ones. As programmers, we find it unfathomable that there could be any value in reducing problems to previously unprogrammed-ones.

To come full circle then, what do I think this has to do with WIA? I believe that WIA is squarely in the middle of these issues, for implementing automata requires both programming and mathematics. WIA is about programming, and about mathematics; it is about the programming of mathematics, and also about the mathematics of programming mathematics. WIA's participants are (we hope) both good programmers and good mathematicians, because we want good systems for doing good mathematics, and we want good mathematics *about* our systems. We cannot live only on one side of the divide between theory and practice; we must straddle it. If we do it well, perhaps we can help to convince the community that the divide need never have been there at all.

Thank you, and please enjoy WIA '96.

Algorithms for Guided Tree Automata

Morten Biehl[1], Nils Klarlund[2], and Theis Rauhe[1]

[1] BRICS, Department of Computer Science,
University of Aarhus,
Ny Munkegade, Aarhus, Denmark
{mbiehl,theis}@brics.dk

[2] AT&T Labs - Research
600 Mountain Ave.,
Murray Hill, NJ 07974
klarlund@research.att.com

Abstract. When reading an input tree, a bottom-up tree automaton is "unaware" of where it is relative to the root. This problem is important to the efficient implementation of decision procedures for the Monadic Second-order Logic (M2L) on finite trees. In [KS97], it is shown how exponential state space blow-ups may occur in common situations. The analysis of the problem leads to the notion of *guided tree automaton* for combatting such explosions. The guided automaton is equipped with separate state spaces that are assigned by a top-down automaton, called the *guide*.

In this paper, we explore the algorithmic and practical problems arising from this relatively complicated automaton concept.

Our solutions are based on a BDD representation of automata [HJJ+96], which allows the practical handling of automata on very large alphabets. In addition, we propose data structures for avoiding the quadratic size of transition tables associated with tree automata.

We formulate and analyze product, projection (subset construction), and minimization algorithms for guided tree automata. We show that our product algorithm for certain languages are asymptotically faster than the usual algorithm that relies on transition tables.

Also, we provide some preliminary experimental results on the use of guided automata vs. standard tree automata.

1 Introduction

The MONA tool [HJJ+96] implements a decision procedure and counter model generator for a Monadic Second-order Logic on strings. This logic has first-order terms that denote positions in the string and limited arithmetic on such terms; in addition, second-order terms denote subsets of positions. The decision procedure follows the classical method of associating a *Deterministic Finite Automaton* (DFA) to each formula in M2L such that the DFA accept the strings satisfying the formula. The main idea is the use of an extended input alphabet such that a string encodes the value of all free variables. Naturally, this encoding leads to very large alphabets, whose representation becomes the major issue in the computational handling of such DFAs.

An extension of MONA to binary trees is currently under development at BRICS in Aarhus. The traditional way of deciding formulas in the Monadic Second-order Logic on Binary Trees is to associate a *Deterministic Finite Tree Automaton* (DFTA) to each subformula. Each automaton represents the set of interpretations that make the subformula true. The automata are calculated according to a simple correspondence between logical connectives and automata-theoretic operations. In [KS97], some sources of exponential and polynomial blow-ups in tree automata associated with M2L formulas are studied. It is shown there that a common source of state space explosion for DFTAs is their lack of knowledge about the position relative to the root of the subtree read by the automaton.

Among the proposals in [KS97], the asymptotically best one is to factorize the state space by a representation called a *guided tree automaton*. The factorization is carried out by a top-down automaton, called the *guide*, which assigns state space to the nodes of the input tree. The transition relation of the bottom-up automaton is thereby split into many components.

In [KS97], a high-level programming language FIDO based on M2L and some conventional programming language concepts is proposed for the convenient expression of properties of parse trees. The compilation of FIDO into M2L is described. Also, a concept of *universe* is proposed as an extension of M2L. A universe declaration defines a separate tree address space. In [KS97], it is argued how universes naturally arise from a FIDO program and how they in turn give rise to guides that reduce state spaces.

The algorithmic aspects of guided tree automata are involved and were not discussed in [KS97].

In the present paper, we propose efficient data structures and algorithms for BDD-represented guided tree automata. A main problem addressed is how to avoid the inherent quadratic blow-up in the representation of a transition relation. This blow-up hinges on the property that for an n-state automaton, there are n^2 pairs for which a function from letters to new states have to be specified. With our solution, we can bound the running time of the product algorithm in certain situations: the total time required is $O(N^{\frac{3}{2}})$, where N is the number of states of the product automaton, whereas a conventional algorithm would use $\theta(N^2)$ time (and space).

Our solution to the quadratic blow-up problem relies on observations that we make about the nature of transition relations occurring during the decision procedure for M2L. We argue that transition relations tend to be *sparse*, at least for the important first-order fragment of M2L. Our representation uses essentially the same default idea as in [CC82], which in our case can be expressed as: for fixed left state q', the largest class of right states q'' such that that the transition function is constant on (q', q'') can be represented by a default transition. All other q'' must be represented by explicit entries for (q', q''). Our contribution is to solve a number of technical problems that must be overcome in order to use this idea efficiently, that is, so that asymptotic and practical gains can be achieved.

The rest of the paper is organized as follows. In Section 2, we provide the definitions of guided tree automata and discuss their relation to usual tree automata. In Section 3, we suggest efficient data structures for the representation of guided tree automata. In Section 4, we give detailed accounts of algorithms for product, projection (including determinization), and minimization. Finally, we report on some preliminary experimental results in Section 5.

2 Guided tree automata

Let Σ be an alphabet. The *trees* T_Σ over Σ are denoted as follows. A *leaf* is identified with the empty tree ε. Internal nodes are identified with their subtrees, which are of the form $\alpha\langle t_1, t_2 \rangle$, where $\alpha \in \Sigma$ is the *label* of the node and $t_1, t_2 \in T_\Sigma$ are the left and right subtrees. Leaves do not have labels. A *Deterministic Finite Tree Automaton* (DFTA) is a tuple $M = (Q, \Sigma, q_0, F, \delta)$, where Q is the finite set of states, $F \subseteq Q$ is the final states, $q_0 \in Q$ is the initial state and δ is the transition function $\delta : (Q \times Q) \to (\Sigma \to Q)$. The labeling function $\hat{\delta} : T_\Sigma \to Q$ is defined inductively by

$$\hat{\delta}(\varepsilon) = q_0$$
$$\hat{\delta}(\alpha\langle t_1, t_2 \rangle) = \delta(\hat{\delta}(t_1), \hat{\delta}(t_2))(\alpha)$$

We say that M *accepts* the *input tree* t if $\hat{\delta}(t) \in F$. Thus informally, the automaton traverses the tree bottom-up while associating a state to each subtree. The set $L(M)$ of all trees accepted is the *language* accepted by M. A tree language is *regular* if and only if it is the language accepted by some DFTA.

According to the definition above, the traversal of a subtree is independent of where the subtree is positioned in the input tree. As argued in [KS97], it would often be beneficial to provide the automaton with "positional knowledge" indicating what part of the input tree, relative to the root, the automaton is currently traversing.

To see how such information may bring about a reduction in the state space, consider the tree language L consisting of all trees for which the number of occurrences of $\alpha \in \Sigma$ in the left subtree is divisible by n and the number of occurrences of $\beta \in \Sigma$ in the right subtree is divisible by m. A DFTA recognizing L then requires $n \cdot m$ states (when n and m are relative primes). The language is more efficiently recognized by means of three separate automata : M_L for traversing the left subtree (n states), M_R for traversing the right subtree (m states) and M_T for combining the results. The polynomial state space explosion ($n \cdot m$) is avoided, because each automaton has "positional knowledge." For example, M_L "knows" that it is traversing the left subtree, and thus it does not need to keep track of occurrences of β.

Technically, "positional knowledge" can be provided by a top-down automaton that identifies regions of the input trees.

Definition 2.1. A *guide* $G = (D, \mu, d_0)$ consists of

> D, a finite set of *state space IDs*,
>
> $\mu : D \to D \times D$, the *guide function*, and
>
> $d_0 \in D$, the *initial ID*.

The *size* γ of G is the cardinality of D.

A simple guide that identifies whether a node is a left child, a right child, or the root of the input tree can be defined as $G = (D, \mu, d_0)$, where

$$D = \{top,\ left,\ right\},$$

$$d_0 = top,\ \text{and}$$

$$\mu(d) = (left,\ right),\ \text{for any } d \in D.$$

A *guided tree automaton* defines a state space for each state space ID and a transition function for each transition of the guide:

Definition 2.2. Let $G = (D, \mu, d_0)$ be a guide. A *guided tree automaton* (GTA) M_G with guide G is of the form $(\{Q_d\}_{d \in D}, \Sigma, \{\delta_d\}_{d \in D}, \{\bar{q}_d\}_{d \in D}, F)$. The components of M_G are as follows.

- $\{Q_d\}_{d \in D}$ is a family of disjoint finite sets of states, one set for each state space ID. We often abbreviate this family $\{Q\}_D$.
- Σ is the alphabet.
- $\{\delta_d\}_{d \in D}$ is a family of transition functions, one for each state space ID, such that if $\mu(d) = (d', d'')$ for some $d, d', d'' \in D$, then δ_d is a transition function of the form $\delta_d : (Q_{d'} \times Q_{d''}) \to (\Sigma \to Q_d)$. We say that δ_d *is of type* $d' \times d'' \to d$, and call d' the *left ID* of δ_d, d'' the *right ID* of δ_d and d the *target ID* of δ_d. Similarly, we refer to $Q_{d'}$ as the *left states* of δ_d, $Q_{d''}$ as the *right states* of δ_d and Q_d as the *range states* of δ_d. We often abbreviate this family $\{\delta\}_D$.
- $\{\bar{q}_d\}_{d \in D}$ is the family of initial states, one for each state space ID. We often abbreviate this family $\{\bar{q}\}_D$.
- $F \subseteq Q_{d_0}$ is the set of final states.

The *size* n of M_G is the cardinality of the largest state space in M_G.

Note that if $G = (D, \mu, d_0)$ is a guide with $|D| = 1$, then any tree automaton guided by G is just an ordinary DFTA.

We will rely on notational shortcuts to improve readability. States and sets of states are usually subscripted by the state space ID, denoted by d, d', d'', etc. Where no confusion arises, we write q instead of q_d, q' instead of $q_{d'}$, q'' instead of $q_{d''}$ etc. Similar abbreviations are used for sets of states.

The intuition behind guided tree automata is quite simple. First, the guide labels each node in the tree with a state space ID ("positional knowledge"). Second, each leaf is labeled with the initial state of the state space indicated by the ID. Then in a bottom-up manner, every remaining internal node is labeled with a state of the space indicated by the ID according to the corresponding transition function.

To make these notions more precise, we let $T_{(\Sigma_1, \Sigma_2)}$ denote the set of trees whose leaves belong to the alphabet Σ_1 and for which all other nodes belong to the alphabet Σ_2. The ID labeling function $\hat{\mu} : T_\Sigma \to T_{(\{\bullet\} \times D, \Sigma \times D)}$ is now defined by $\hat{\mu}(t) = \tilde{\mu}(t, d_0)$, where

$$\tilde{\mu}(\varepsilon, d) = (\bullet, d),$$
$$\tilde{\mu}(\alpha\langle t', t''\rangle, d) = (\alpha, d)\langle \tilde{\mu}(t', d'), \tilde{\mu}(t'', d'')\rangle \text{ where } \mu(d) = (d', d'')$$

and $\bullet$ is a symbol not in Σ. The state labeling function $\hat{\delta} : T_{(\{\bullet\} \times D, \Sigma \times D)} \to \bigcup\{Q\}_D$ is defined by:

$$\hat{\delta}(\bullet, d) = \bar{q}_d \text{ and}$$
$$\hat{\delta}((\alpha, d)\langle(t', d'), (t'', d'')\rangle) = \delta_d(\hat{\delta}(t', d'), \hat{\delta}(t'', d''))(\alpha),$$

where δ_d is of type $d' \times d'' \to d$. Note that since $\hat{\mu}$ attaches d_0 to the root of a tree t, we have $\hat{\delta} \circ \hat{\mu}(t) \in Q_{d_0}$. A tree $t \in T_\Sigma$ is said to be accepted by a guided tree automata M_G if $\hat{\delta} \circ \hat{\mu}(t) \in F$, and the set of trees accepted by M_G is denoted $L(M_G)$.

The following proposition constructively shows how DFTAs can be simulated by GTAs and vice versa.

Proposition 2.3.

(a) *Let* $M = (Q, \Sigma, \delta, q_0, F)$ *be a DFTA of size* n *and* $G = (D, \mu, d_0)$ *a guide. Then there is a GTA* $M_G = (\{P\}_D, \Sigma, \{\gamma\}_D, \{\bar{p}\}_D, E)$ *of size* n *such that* $L(M) = L(M_G)$.

(b) *Let* $M_G = (\{Q\}_D, \Sigma, \{\delta\}_D, \{\bar{q}\}_D, F)$ *be a GTA of size* n *guided by* $G = (D, \mu, d_0)$ *of size* γ. *Then there is a DFTA* $M = (P, \Sigma, \gamma, p_0, E)$ *of size* n^γ *such that* $L(M_G) = L(M)$.

Proof. (Idea)

(a) We define M_G by making a copy of the state space and transition function for each $d \in D$.

(b) M simulates all the transition functions of M_G in parallel.

3   Data structures for guided automata

Decision procedures and large alphabets The decision procedure for M2L on strings associates a language over an alphabet of the form $\Sigma = \mathbb{B}^k$ (where $\mathbb{B} = \{0, 1\}$ is the Booleans) to each formula ϕ, see [Tho90]. The idea is the following: for a word $w \in (\mathbb{B}^k)^*$ of length ℓ, each of the k components defines a bit pattern or *track* of length ℓ. Each free variable of ϕ is assigned to a track and is interpreted as the subset of positions in $\{0, \dots, \ell - 1\}$ for which the track contains a 1. The language defined for ϕ is the set of strings that interpret ϕ to be true. Since the number of free variables can be large, say 100, the resulting alphabets can be of astronomical size, say 2_{100}.

Shared BDD representation In the MONA implementation of automata on finite strings [HJJ+96], a shared multi-terminal BDD, called the Σ-*BDD* is used to represent the transition function.

BDDs were originally introduced in [Bry86]. We use the variety defined as follows.

Notation A *Binary Decision Diagram* (BDD) is a rooted, directed graph. Each node $\omega \in \Omega$ is either an *internal node* or a *leaf*. A leaf ω defines a *leaf value* in V, where V is a finite set. An internal node ω possesses an *index* together with a *low successor* and a *high successor* such that the index of both successors is higher than the index of ω. Each node ω represents a function $\mathbb{B}^k \to V$ for some k, and we use ω to denote both the node and the function it represents. Thus, if ω is a root and **b** is a vector of k bits, then we denote the value of the function on **b** by $\omega(\mathbf{b})$.

The kind of BDD defined above is sometimes called a *multi-terminal* BDD. A *shared* BDD is a BDD with multiple roots. In the algorithms in the following sections, we use the *apply* and *restrict* operations described in [Bry86], although we use the term *projection* in place of restriction.

If ω and ω' are roots, then we denote by $\omega * \omega'$ the pairing of functions ω and ω', that is, $\omega * \omega'(\mathbf{b}) = (\omega(\mathbf{b}), \omega'(\mathbf{b}))$. This function can be calculated by a binary BDD apply operation in time bounded by the product of the sizes of ω and ω'.

In our automaton representation, each state of the automaton points to a BDD node, and each leaf value is a state. Given a letter $\alpha \in \mathbb{B}^k$, we may find the value of the transition function by following the BDD nodes according to α from the node pointed to by the state. The leaf reached contains the name of the next state. A MONA representation of a DFA is depicted in Figure 1.

The BDD-based representation of automata on strings allows efficient implementation of standard operations on finite automata (except for minimization, where our current algorithm is quadratic although its behavior in practice is better than quadratic).

BDD-based tree automaton data structure We would like to use a similar representation to gain efficient algorithms for guided tree automata. Thus, we assume that a shared BDD is used for representing the alphabetic part of each transition relation, i.e. the part with signature $\Sigma \to Q$ of $\delta_d : (Q' \times Q'') \to (\Sigma \to Q)$. This BDD is called the Σ-*BDD* and we call the function $\Sigma \to Q$ that it represents the Σ-*behavior*. We use Ω to denote the nodes of the Σ-BDD.

A naïve approach for representing $\delta_d : (Q' \times Q'') \to (\Sigma \to Q)$ is to create an entry for each pair $\langle q', q'' \rangle \in Q' \times Q''$. An entry defines a σ-behavior as a BDD node in a Σ-BDD. This approach leads to an unfortunate quadratic growth, since each of the $|Q'| \times |Q''|$ entries must be explicitly represented.

An alternative approach is to define a binary encoding of the state spaces Q' and Q''. Each state $q'' \in Q''$ has a unique identifier in the range $\{0, \ldots |Q''|-1\}$,

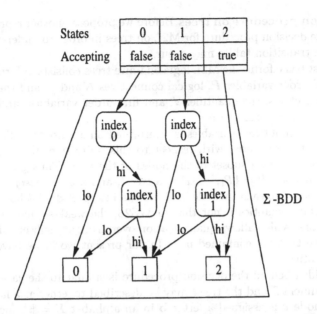

States	0	1	2
Accepting	false	false	true

Figure 1. MONA representation of DFA that accepts all strings over $\mathbb{B}^2$ with at least two occurrences of the letter "11".

which can be encoded by means of a vector of $k = \log(|Q''|)$ bits. Thus for each $q' \in Q'$, we can define a binary function

$$f_{q'}(b_0, \ldots, b_{k-1}) = \delta_d(q', q''),$$

where $b_0, \ldots, b_{k-1}$ is the binary encoding of q''. All these functions can now be represented in a shared BDD. Furthermore, we still use a shared BDD for representing the Σ-behavior $\delta_d(q', q'')$.

For a fixed q', this representation represents the functions $\delta_d(q', q'')$, where q'' ranges over all Q'', succinctly in the case that $\delta_d(q', q'')$ has the same value, which we call a *default Σ-behavior*, for almost all q''. For instance, if 99% of all q'' lead to the same Σ-BDD node, then the state BDD for q' has approximately $0.01 \cdot n$ nodes (when $.01 \cdot n >> \log n$). The total representation is then only one hundredth the size of a total transition table. Thus our notion of *sparsity* is that for any left state, very few right states are of interest—the rest all lead to the default behavior.

An experimental version of the extended MONA system has been implemented based on this representation with acceptable, although not astonishing results. Since the state encoding tends to be random, the compression occurs only for the situations just described. Also, there is a $O(\log n)$ penalty for looking up the transition function for a particular q'', since $O(\log n)$ BDD nodes must be followed. Our experience is that this factor creeps into automata algorithms based on this representation. Therefore, we will pursue in this paper a representation without a logarithmic overhead.

M2L decision procedure on trees Before we propose another representation, we review the decision procedure for M2L on trees in order to understand where sparseness in transition tables may occur.

In its most basic form, the M2L logic on finite trees consists of formulas made out of second order variables P, logical connectives $\wedge$ and $\neg$, and the quantifier $\exists$. Other connectives, the quantifier $\forall$, and first-order variables can be reduced to expressions in the basic form.

A formula without free variables is interpreted on a finite tree. Second-order quantification is interpreted with respect to this finite tree, that is, a second-order variable denotes a subset of the nodes of the tree. Thus given a formula ϕ with free variables $\mathcal{P} = \{P_1, \cdots, P_k\}$, a tree t, and a *value assignment* $\mathcal{I}$ that assigns subsets of t to each variable in $\mathcal{P}$, we can interpret ϕ. If ϕ holds, we write $t, \mathcal{I} \models \phi$. (This semantics is not that of WS2S, the weak-second order theory of two succesors, which allows quantification over arbitrary subsets of positions, not only those that are contained in t. In our present work, we have used this simpler semantics.)

The key idea behind the decision procedure is similar to the case of strings: a value assignment $\mathcal{I}$ and the tree t may be described together as a labeled tree, where each node v is assigned a letter **b** in an alphabet $\Sigma = \mathbb{B}^k$ such that the ith component of **b** denotes whether v is in P_i. We denote this labeled tree by $(t, \mathcal{I})$. The P_i-*track* of $(t, \mathcal{I})$ is the $\mathbb{B}$-labeled tree that is gotten by restricting the labeling on t to the ith component.

By structural induction on formulas, we construct automata $A^{\phi, \mathcal{P}}$ on alphabet $\mathbb{B}^k$, where $k = |\mathcal{P}|$, satisfying the correspondence:

$$t, \mathcal{I} \models \phi \text{ iff } (t, \mathcal{I}) \in L(A^{\phi, \mathcal{P}})$$

Thus, $A^{\phi, \mathcal{P}}$ accepts exactly the labeled trees $(t, \mathcal{I})$ that make ϕ true.

Sparsity of M2L transition relations Typically, there are not many $\mathcal{I}$ that make sense due to the many constraints that are imposed on variables; for example, if the second-order variable P is used in ϕ to denote a single position (corresponding to the common case where quantification is first-order), then ϕ is trivially false if there are more than one position in the P-track containing a 1. Once a second such position is encountered by the automaton in its bottom-up parsing of the tree, it will go to a "reject-all state" (a graph-theoretic sink). So, intuitively, each state will contain information or *assumptions* specifying the "first-order status" of P, namely whether 0, 1, or more occurrences of a 1 have been encountered. Thus, if we consider a random left state q' and a random right state q'', then a transition to a state other than the "reject-all state" can happen only if q' and q'' for each variable make consistent assumptions about the its status. For example, in the case of a second-order variable P modeling a first-order variable, the "reject-all state" will result from any scenario where each of q' and q'' contains the first-order status assumption that a single 1 in the P-track has already occurred. Therefore, given k first-order variables, the chance that a random pair of states q' and q'' are consistent is ρ^k, where $\rho < 1$.

(Here, we have assumed a uniform probability distribution and that the first-order status information is not masked by other information; thus, the number of states is also exponential in k.) This argument could be formalized so as to show that sparsity occurs under some rather representative circumstances when M2L is translated to tree automata.

Our representation Our representation exploits the assumed existence of a preponderant equivalence class by storing it implicitly in a manner similar to the incompletely specified transition functions of [CC82].

To make notions more precise, consider a transition function δ_d of type $d' \times d'' \to d$. A state $q' \in Q'$ induces an equivalence relation $\equiv_{q', \delta_d}$ on Q'' defined by

$$q_1'' \equiv_{q', \delta_d} q_2'' \quad \text{iff} \quad \delta_d(q', q_1'')(\alpha) = \delta_d(q', q_2'')(\alpha) \quad \text{for all } \alpha \in \Sigma.$$

Sparsity means that one equivalence classes of $\equiv_{q', \delta_d}$ contains almost all of Q'', whereas the other equivalence classes only contain a few elements of Q'' each. An equivalence class that contains at least as many elements as any other equivalence class is referred to as a *largest* equivalence class.

In our representation, we store in an array the following information for each of the left states $q' \in Q'$:

- a node in the Σ-BDD, named $q'.default_d$, denoting the default behavior, and
- a set of pairs $(q'', \omega) \in Q'' \times \Omega$, named $q'.explicit_d$.

The subscript d indicates that the associations are with respect to the transition function with target ID d, i.e. the above associations are made for each transition function. Note that this representation is asymmetric in the sense that $default_d$ and $explicit_d$ are defined only for left states. (The choice of left is arbitrary.)

We represent a largest equivalence class $[q_*'']$ of $\equiv_{q', \delta_d}$ implicitly by setting $q'.default_d = \delta_d(q', q_*'')$. For all $q'' \notin [q_*'']$, the pair $(q'', \delta_d(q', q''))$ is stored in $q'.explicit_d$. Thus for arbitrary $q'' \in Q''$ and $\alpha \in \Sigma$, we have

$$\delta_d(q', q'')(\alpha) = \begin{cases} \omega(\alpha) & \text{if } (q'', \omega) \in q'.explicit_d \text{ for some } \omega \\ q'.default_d(\alpha) & \text{otherwise} \end{cases}$$

Thus determining $\delta_d(q', q'')$ amounts to a set lookup. The representation of a transition function is depicted in Figure 2.

Representing the whole GTA A GTA $M_G = (\{Q\}_D, \Sigma, \Delta, \{\bar{q}\}_D, F)$ is implemented by a structure as described above for each transition function. We use a single BDD, shared among all transition functions, to encode Σ. In addition, the implementation has a bit vector of size $|Q_{d_0}|$ to represent F.

4 Algorithms for automata operations

We show how algorithms for the four basic operations on automata, namely complementation, product, projection (and subsequent subset construction for determinization), and minimization can be formulated in terms of our representation for guided tree automata.

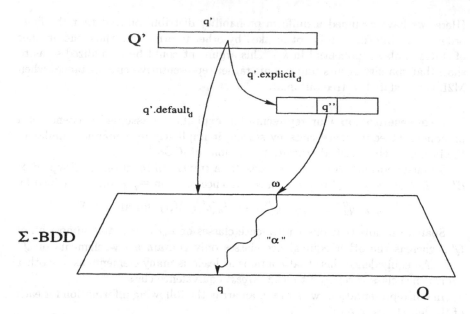

Figure 2. Representation of a transition function δ_d of type $d' \times d'' \rightarrow d$.

4.1 Complementation

The operation of complementation is easy to carry out: simply negate the bit vector representing the final states.

4.2 Product

Let $\mathfrak{P} = (\{P\}_D, \Sigma, \{\gamma\}_D, \{\bar{p}\}_D, E)$ and $\mathfrak{Q} = (\{Q\}_D, \Sigma, \{\phi\}_D, \{\bar{q}\}_D, F)$ be tree automata guided by $G = (D, \mu, d_0)$. The product automaton of $\mathfrak{P}$ and $\mathfrak{Q}$ is then $\mathfrak{R} = (\{R\}_D, \Sigma, \{\delta\}_D, \{\bar{r}\}_D, H)$, where

$$R_d = P_d \times Q_d$$
$$\delta_d = \gamma_d \times \phi_d : ((P'_{d'} \times Q'_{d'}) \times (P''_{d''} \times Q''_{d''})) \rightarrow (\Sigma \rightarrow (P_d \times Q_d)), \text{where}$$
$$\gamma_d \times \phi_d(\langle p', q'\rangle, \langle p'', q''\rangle)(\alpha) = \langle \gamma_d(p', p'')(\alpha), \phi_d(q', q'')(\alpha)\rangle$$
$$\bar{r}_d = \langle \bar{p}_d, \bar{q}_d\rangle$$
$$H = E \times F$$

A relatively simple approach We begin by describing a simple approach of performing a product of two guided tree automata. This approach does not construct the whole product spaces $P_d \times Q_d$; instead, only reachable pairs of states are calculated.

The algorithm maintains two sets for each $d \in D$:

- R_d is the set of *reachable pairs* in $P_d \times Q_d$ encountered at any stage of the computation. The set is initialized to $\{\langle \bar{p}_d, \bar{q}_d \rangle\}$.
- *unprocessed$_d$* is the subset of the reachable pairs R_d for which the transition function has not been calculated. Initially, *unprocessed$_d$* is also the singleton set $\{\langle \bar{p}_d, \bar{q}_d \rangle\}$.

The transition function to be calculated of type $d' \times d'' \to d$ is represented by explicit sets of the form $\langle p', q' \rangle.explicit_d$ and default values $\langle p', q' \rangle.default_d$.

Consider a pair $\langle \hat{p}, \hat{q} \rangle$ from some set *unprocessed$_{\hat{d}}$*. For every transition function δ_d of type $\hat{d} \times d'' \to d$, we must calculate all transitions involving $\langle \hat{p}, \hat{q} \rangle$. We say that we process $\langle \hat{p}, \hat{q} \rangle$ under the *left view*. Specifically for each $\langle p'', q'' \rangle \in R_{d''}$, we calculate

$$\omega = \gamma_d(\hat{p}, p'') * \phi_d(\hat{q}, q'')$$

by performing a binary BDD apply. Since ω equals $\delta_d(\langle \hat{p}, \hat{q} \rangle, \langle p'', q'' \rangle)$, we can extend δ_d for this value by inserting $(\langle p'', q'' \rangle, \omega)$ in $\langle \hat{p}, \hat{q} \rangle.explicit_d$.

We also have to process under the *right view*: for transition functions of type $d' \times \hat{d} \to d$, we compute for each pair $\langle p', q' \rangle \in R_{d'}$

$$\omega = \gamma_d(p', \hat{p}) * \phi_d(q', \hat{q})$$

and insert $(\langle \hat{p}, \hat{q} \rangle, \omega)$ into $\langle p', q' \rangle.explicit_d$.

After all transitions where $\langle \hat{p}, \hat{q} \rangle$ occurs have been considered, $\langle \hat{p}, \hat{q} \rangle$ is removed from *unprocessed$_d$*.

During computation of each BDD apply, all pairs that occur as BDD leaf values and that have not been encountered before are inserted into *unprocessed$_d$* and R_d.

The algorithm continues until every set *unprocessed$_d$* is empty. It is easy to verify that the algorithm maintains the invariant that for all $d \in D$, the transition functions δ_d are totally defined for $(R_{d'} \setminus unprocessed_{d'}) \times (R_{d''} \setminus unprocessed_{d''})$. Hence upon termination, all transition functions are defined (through the sets *explicit$_d$*) for all reachable pairs.

To compress the representation of the state part of the transition relation, an extra sweep upon termination is necessary to adjust the *default$_d$* values. Alternatively, this can be done on-line by keeping track of the frequency of BDD nodes inserted in the *explicit$_d$* sets. In this way, space cost is decreased.

To analyze the above algorithm, assume for simplicity that the number of reachable states in any state space of the resulting automaton is bounded by N, i.e. $|R_d| \leq N$ for all d. The time used per state in the resulting automaton to is $O(N)$, and hence in total the algorithm uses time $O(N^2)$. (In this paper, we ignore the size of the σ-BDDs in our complexity estimates.)

A more efficient approach The algorithm we propose in this paper is designed to take advantage of the *default$_d$* values of the automata $\mathfrak{P}$ and $\mathfrak{Q}$. In certain cases, this method makes it possible to obtain an $O(N^{\frac{3}{2}})$ time bound in contrast to the quadratic bound above.

The main idea explained for the left view is: when processing a pair $\langle \hat{p}, \hat{q} \rangle$ for a transition function δ_d of type $\hat{d} \times d'' \to d$, we can often avoid considering all pairs $\langle p'', q'' \rangle$ of state space $R_{d''}$. This is accomplished by temporarily letting $\langle \hat{p}, \hat{q} \rangle . default$ equal $\hat{p}.default_d * \hat{q}.default_d$. If

$$\text{either } p'' \in \hat{p}.explicit_d \text{ or } q'' \in \hat{q}.explicit_d, \tag{1}$$

then we do have to consider $\langle p'', q'' \rangle$; otherwise, when (1) does not hold, it is the case that $\delta_d(\langle \hat{p}, \hat{q} \rangle, \langle p'', q'' \rangle) = \langle \hat{p}, \hat{q} \rangle . default$, and hence we do not need to insert $\langle p'', q'' \rangle$ into the set $\langle \hat{p}, \hat{q} \rangle . explicit_d$.

Similarly, under the right view, for transition function δ_d of type $d' \times \hat{d} \to d$, we need to insert $\langle \hat{p}, \hat{q} \rangle$ in $\langle p', q' \rangle . explicit_d$ only if

$$\text{either } \hat{p} \in p'.explicit_d \text{ or } \hat{q} \in q'.explicit_d. \tag{2}$$

Since the purpose of the efficient approach is to avoid considering data implied by the default behaviors, we need to introduce additional data structures in which the pairs that need consideration can be explicitly looked up.

To do this, we need to precompute some information for each transition function δ of type $d' \times d'' \to d$. For automaton $\mathfrak{P}$, the following is computed:

- To each state $p' \in P_{d'}$, we define

$$p' \to expl_d = \{p'' \mid (p'', \omega) \in p'.explicit\}.$$

These sets are calculated in order to maintain an explicit representation of the pairs that satisfy (1).
- To each state $p'' \in P_{d''}$ we let

$$p'' \leftarrow expl_d = \{p' \mid (p'', \omega) \in p'.explicit\}.$$

These sets are for the computational handling of requirement (2).

Similar information is calculated for $\mathfrak{Q}$. These calculations can be carried out in linear time of the size of the representation.

In addition to the sets of reachable pairs R_d and the sets *unprocessed*, the algorithm maintains certain *explicit pairs sets*, which are subsets of the reachable pairs sets. For automaton $\mathfrak{P}$, these sets are

- $R|_{p' \to expl_d} = \{\langle p'', q'' \rangle \in R_{d''} \mid p'' \in p' \to expl_d\}$ and
- $R|_{p'' \leftarrow expl_d} = \{\langle p', q' \rangle \in R_{d'} \mid p' \in p'' \leftarrow expl_d\}$

for all $p' \in P_{d'}$ and $p'' \in P_{d''}$ where $\mu(d) = (d', d'')$. Similar sets are defined for automaton $\mathfrak{Q}$.

The sets $R|_{p' \to expl_d}$ determine whether a pair $\langle p'', q'' \rangle$ belongs to the *explicit* list of $\langle \hat{p}, \hat{q} \rangle$ under the left view according to (1): if $\langle p'', q'' \rangle \in R|_{\hat{p} \to expl_d} \cup R|_{\hat{q} \to expl_d}$, then p'' is found in $\hat{p}.explicit_d$ or q'' is found in $\hat{q}.explicit_d$. In either case, it the *explicit* set for $\langle \hat{p}, \hat{q} \rangle$ must contain $\langle p'', q'' \rangle$ (unless the Σ-behavior of $(\langle \hat{p}, \hat{q} \rangle, \langle p'', q'' \rangle)$ happens to be the default $\hat{p}.default_d * \hat{q}.default_d$). Therefore, the processing of pair $\langle \hat{p}, \hat{q} \rangle$ under the left view need to involve only right states $\langle p'', q'' \rangle \in R|_{\hat{p} \to expl_d} \cup R|_{\hat{q} \to expl_d}$.

The sets $R|_{p'' \leftarrow expl_d}$ play a similar role for the processing under the right view.

Figure 3 shows how the *unprocessed* sets and the various versions of the sets of reachable states are extended as a result of new pairs that are generated during an apply operation:

```
fun apply_and_extend(ω₁, ω₂, d̂) =
    use the BDD apply operation on ω₁ and ω₂ and for each
    pair of states ⟨p̂, q̂⟩ encountered in a pair of leaves and not
    already in either R_d̂ ∪ unprocessed_d̂ do the following:
        add ⟨p̂, q̂⟩ to unprocessed_d̂
        for d̂ × d'' → d for some d, d'' do
            add ⟨p̂, q̂⟩ to R|_{p''←expl_d} for each p'' ∈ p̂ → expl_d
            add ⟨p̂, q̂⟩ to R|_{q''←expl_d} for each q'' ∈ q̂ → expl_d
        od
        for d' × d̂ → d for some d, d' do
            add ⟨p̂, q̂⟩ to R|_{p'→expl_d} for each p' ∈ p̂ ← expl_d
            add ⟨p̂, q̂⟩ to R|_{q'→expl_d} for each q' ∈ q̂ ← expl_d
        od
    return the resulting BDD
```

Figure 3. Algorithm for auxiliary function *apply_and_extend*.

Figure 4 summarizes how the main part of the product algorithm works. After the product automaton has been calculated, the product algorithm must also ensure that the default transition in each case corresponds to an equivalence class of maximal size. If not, a maximal equivalence class will be converted from an explicit representation to a default representation and the former default representation is made explicit. These calculations can be carried out in time linear to the product automaton.

Analysis Consider the case where all state spaces of $\mathfrak{P}$ and $\mathfrak{Q}$ are bounded by n. In addition let $t(n)$ be a function of n that bounds the number of elements in any set $explicit_d$ associated to any state in $\mathfrak{P}$ or $\mathfrak{Q}$. We then claim that our algorithm uses at most time $O(t(n)n)$ per state in the product automaton $\mathfrak{R}$. Consider the pair $\langle \hat{p}, \hat{q} \rangle$. As discussed in the previous section, the transition function δ_d of type $d' \times d'' \to d$ is extended under the left view for only those pairs $\langle p'', q'' \rangle$

$R_d = \emptyset$ for all $d \in D$
$unprocessed_d = \{\langle \bar{p}_d, \bar{q}_d \rangle\}$ for all $d \in D$
while $unprocessed_{\hat{d}} \neq \emptyset$ for some $\hat{d} \in D$ **do**
    remove a pair $\langle \hat{p}, \hat{q} \rangle$ from $unprocessed_{\hat{d}}$
    for $\hat{d} \times d'' \to d$ for some d, d'' **do**
        $\langle \hat{p}, \hat{q} \rangle.default_d = apply\_and\_extend(\hat{p}.default_d, \hat{q}.default_d, d)$
        for all $\langle p'', q'' \rangle \in R|_{\hat{p} \to expl_d} \cup R|_{\hat{q} \to expl_d}$ **do**
            $\omega = apply\_and\_extend(\gamma_d(\hat{p}, p''), \phi_d(\hat{q}, q''), d)$
            if $\omega \neq \langle \hat{p}, \hat{q} \rangle.default_d$ **then**
                add $(\langle p'', q'' \rangle, \omega)$ to $\langle \hat{p}, \hat{q} \rangle.explicit_d$
            fi
        od
    od
    for $d' \times \hat{d} \to d$ for some d, d' **do**
        for all $\langle p', q' \rangle \in R|_{\hat{p} \leftarrow expl_d} \cup R|_{\hat{q} \leftarrow expl_d}$ **do**
        $\langle p', q' \rangle.default_d = apply\_and\_extend(p'.default_d, q'.default_d, d)$
        $\omega = apply\_and\_extend(\gamma_d(p', \hat{p}), \phi_d(q', \hat{q}), d)$
        if $\omega \neq \langle p', q' \rangle.default_d$ **then**
            add $(\langle \hat{p}, \hat{q} \rangle, \omega)$ to $\langle p', q' \rangle.explicit_d$
        fi
        od
    od
    add $\langle \hat{p}, \hat{q} \rangle$ to state space $R_{\hat{d}}$
od

Figure 4. Algorithm for product construction for GTAs.

from $R_{d''}$ for which either $p'' \in \hat{p}.explicit_d$ or $q'' \in \hat{q}.explicit_d$. The number of such pairs is bounded by $|\hat{p}.explicit_d| * |Q_{d''}| + |\hat{q}.explicit_d| * |P_{d''}| = O(t(n) \cdot n)$. (Similar considerations apply to the calculations done under the right view.) In contrast, the simple algorithm visits all reachable pairs, so it uses time $O(n^2)$ per product automaton state when all states are reachable. Hence for $t(n) < o(n)$, our time of $O(n \cdot t(n))$ is asymptotically better.

To put this in a sharper light, assume $t(n) = O(1)$. The resulting product automata have size $N = n^2$, and the simple algorithm would be of time complexity $O(N^2)$; in contrast, our algorithm uses time $O(t(n) \cdot n)$ per state of the resulting automata, e.g. it uses total time $O(n \cdot t(n) \cdot n^2) = O(n^3) = O(N^{\frac{3}{2}})$.

4.3 Projection and determinization

Projection of track i, $1 \leq i \leq k + 1$, is the process of converting a guided tree automaton M over $\mathbb{B}^{k+1}$ to a nondeterministic guided tree automaton M' over $\mathbb{B}^k$ by removing information in track i. Thus, $L(M')$ consists of all trees that can be obtained by erasing the ith component of every node label in a tree in $L(M)$. The automaton M' is constructed by applying the BDD projection operation on the Σ-BDD associated with each transition function. The leaf function of the apply operation combines the jointly reachable states into a set that is either a singleton or has two elements. Thus the leaf values of the Σ-BDDs are now sets of states.

Determinization of a nondeterministic GTA is done according to the subset construction. Our algorithm reuses the properties of the *explicit* and *default* values of the product operation. Thus during the subset construction, it is possible to predict that certain subsets of states inherit a default behavior and hence can be avoided during construction of the transition functions. Details will be given in a full version of this paper.

4.4 Minimization

Minimizing guided tree automata is a rather complex task compared to the minimization of ordinary tree automata (which is already a non-trivial affair that as far as we know has not been described in the literature from an algorithmic point of view; but see [Koz92] for an elegant proof that a minimum automaton exists). Before discussing the minimization process, we extend the notation provided by [PT87]:

Notation A *partition* $\mathcal{P}$ of a finite set U is a set of disjoint subsets of U such that the union of these sets is all of U. The elements of a partition are called its *blocks*. A *refinement* $\mathcal{Q}$ of $\mathcal{P}$ is a partition such that any block of $\mathcal{Q}$ is a subset of a block of $\mathcal{P}$. We let $[q]_\mathcal{P}$ denote the block of the partition $\mathcal{P}$ containing the element q, and when no confusion arises, we drop the subscript.

Let $M_G = (\{Q\}_D, \Sigma, \{\delta\}_D, \{\bar{q}\}_D, F)$ be a GTA guided by $G = (D, \mu, d_0)$, and let $\{\mathcal{P}_d\}_{d \in D}$ be a family of partitions such that $\mathcal{P}_d$ is a partition of Q_d. We extend the shorthand notation introduced in the previous sections, and we write $\mathcal{P}$ for the partition $\mathcal{P}_d$, $\mathcal{P}'$ for $\mathcal{P}_{d'}$ etc. when no confusion occurs. Let $\{\mathcal{Q}\}_D$ be a refinement of $\{\mathcal{P}\}_D$, i.e. $\mathcal{Q}_d$ is a refinement of $\mathcal{P}_d$ for all $d \in D$. Let $\delta_d \in \{\delta\}_D$ be a transition function of type $d' \times d'' \to d$.

A block B' of $\mathcal{Q}'$ δ_d-*respects* $\mathcal{P}_d$ if

$$\forall q_1', q_2' \in B', \forall q'' \in Q'', \forall \alpha \in \Sigma : [\delta_d(q_1', q'')(\alpha)]_{\mathcal{P}_d} = [\delta_d(q_2', q'')(\alpha)]_{\mathcal{P}_d}$$

Similarly a block B'' of $\mathcal{Q}''$ δ_d-*respects* $\mathcal{P}_d$ if

$$\forall q_1'', q_2'' \in B'', \forall q' \in Q', \forall \alpha \in \Sigma : [\delta_d(q', q_1'')(\alpha)]_{\mathcal{P}_d} = [\delta_d(q', q_2'')(\alpha)]_{\mathcal{P}_d}$$

Thus B' δ_d-respects $\mathcal{P}_d$ if δ_d cannot distinguish between the elements in B' relative to $\mathcal{P}_d$. A partition $\mathcal{Q}'$ δ_d-respects $\mathcal{P}_d$ if every block of $\mathcal{Q}'$ δ_d-respects $\mathcal{P}_d$, and a family of partitions $\{\mathcal{Q}\}_D$ δ_d-respects $\mathcal{P}_d$ if $\mathcal{Q}'$ and $\mathcal{Q}''$ δ_d-respects $\mathcal{P}_d$. A family of partitions $\{\mathcal{Q}\}_D$ respects the family of partitions $\{\mathcal{P}\}_D$ if $\{\mathcal{Q}\}_D$ δ_d-respects $\mathcal{P}_d$ for all transition functions $\delta_d \in \{\delta\}_D$, where δ_d is of type $d' \times d'' \to d$. A family of partitions is *stable* if it respects itself. The *coarsest, stable family of partitions* $\mathcal{Q}_D$ respecting $\mathcal{P}_D$ is a unique family of partitions such that any other stable family of partitions respecting $\mathcal{P}_D$ is a refinement of $\mathcal{Q}_D$.

The minimization algorithm works by gradually refining a current family of partitions so that each step of the algorithm ensures that the refinement δ_d-respects the current family of partitions for some transition function δ_d. We first show how to split a current family of partitions with respect to a single transition function, and later how this is used to minimize a guided tree automata. We assume for the rest of this section that our representation is symmetric, that this, the sets $explicit_d$ and the default behaviors $default_d$ are also present for right states. It is straightforward to precompute these values from the $explicit_d$ and $default_d$ information of the left states (this is the information calculated by the product and project algorithms).

Splitting with respect to δ_d of type $d' \times d'' \rightarrow d$ Let Q, Q' and Q'' denote the current partition of Q, Q' and Q'' respectively and assume that the current family of partitions does not δ_d-respect Q. We now show how to compute the coarsest partition which δ_d-respects the current partition.

1. Replace the leaf-values in the Σ-BDD by canonical representatives according to Q and reduce it. This induces a partition of the nodes in the Σ-BDD denoted S.
2. Refine Q' to $\mathcal{P}'$ such that $q_1' \equiv_{\mathcal{P}'} q_2'$ iff $q_1' \equiv_{Q'} q_2'$ and $\forall q'' \in Q''$, $\delta_d(q_1', q'') \equiv_S \delta_d(q_2', q'')$
3. Refine Q'' to $\mathcal{P}''$ such that $q_1'' \equiv_{\mathcal{P}''} q_2''$ iff $q_1'' \equiv_{Q''} q_2''$ and $\forall q' \in Q'$, $\delta_d(q', q_1'') \equiv_S \delta_d(q'q_2'')$

Step 1 ensures $\omega \equiv_S \omega'$ iff $\forall \alpha \in \Sigma$ $\omega(\alpha) \equiv_Q \omega'(\alpha)$. For the partition calculated in step 2 we have $q_1' \equiv_{\mathcal{P}'} q_2'$ iff $\delta_d(q_1', q'') \equiv_S \delta_d(q_2', q'')$ for all $q'' \in Q''$, i.e. $\forall q'' \in Q''$, $\forall \alpha \in \Sigma$, $\delta_d(q_1', q'')(\alpha) \equiv_Q \delta_d(q_2', q'')(\alpha)$. Thus all blocks of $\mathcal{P}'$ δ_d-respect Q. Similarly, step 3 ensures that all blocks of $\mathcal{P}''$ δ_d-respect Q.

The refinement operation in step 2 is performed by assigning to each element $q' \in Q'$ a canonical representative for its block in the new partition $\mathcal{P}'$ respecting Q. Similar representatives are calculated for $q'' \in Q''$ in step 3. For a Σ-BDD node ω, we denote its canonical representative with respect to S by $\hat{\omega}$.

We now address the problem of calculating the canonical representatives in step 2. (Step 3 is symmetric.) Consider a state $q' \in Q'$. The problem is to calculate in linear time a unique characterization of the function $q'' \mapsto [\delta_d(q', q'')]$. We must deal with the default representation while making sure that the characterization remains unique. The following techniques allow a default based representation, where the default case is used only when its uniqueness can be assured.

Let $\nu_{q'} = q'.default_d$ and let $\hat{\nu}_{q'}$ be its representative according to S. By traversing the states in $q'.explicit_d$, we calculate the set:

$$M_{q'} = \{(q'', \hat{\omega}) \mid (q'', \omega) \in q'.explicit_d \text{ and } \hat{\omega} \neq \hat{\nu}_{q'}\}.$$

If $|M_{q'}| < \frac{1}{2}|Q''|$, then $\hat{\nu}_{q'}$ is a default behavior that applies to more than half the states in Q''. Otherwise, $|M_{q'}| \geq \frac{1}{2}|Q''|$ and we find a $\hat{\nu}_{q'}'$ minimizing the size of

$$\{(q'', \hat{\omega}) \mid (q'', \omega) \in Q'' \times \Omega, \ \omega = \delta_d(q', q'') \text{ and } \hat{\omega} \neq \hat{\nu}'_{q'}\},$$

by another linear traversal. Redefine $M_{q'}$ to be this set, and $\hat{\nu}_{q'}$ to be $\hat{\nu}'_{q'}$. If the size of $|M_{q'}|$ still is larger than $\frac{1}{2}|Q''|$, then there is no way of characterizing the default behavior uniquely by means of the class that contains more than half the states. Thus in this case, we redefine $M_{q'}$ once more to be:

$$\{(q'', \hat{\omega}) \mid (q'', \omega) \in Q'' \times \Omega, \ \omega = \delta_d(q', q'')\},$$

by utilizing all states in Q'' and redefine $\hat{\nu}_{q'}$ to be a fixed value $\perp$ different from any representative $\hat{\omega}$. We remember the old values of $M_{q'}$ and $\hat{\nu}_{q'}$ and denote these as $M_{q'}^{\perp}$ and $\hat{\nu}_{q'}^{\perp}$ respectively. It is now not difficult to show that the tuple $(M_{q'}, \hat{\nu}_{q'})$ is a canonical representative for the block in $\mathcal{P}'$ containing q', i.e. $q'_1 \equiv_{\mathcal{P}'} q'_2$ iff $(M_{q'_1}, \hat{\nu}_{q_1}) = (M_{q'_2}, \hat{\nu}_{q_2})$. Also, it can be seen that the calculation of the representative is linear in the size of $q'.explicit_d$. We note that in practice we would additionally need to calculate a canonical index (an integer) from the canonical representative using some hashing approach.

In total, calculating the canonical representative for a state q' in step 2 takes time $O(|q'.explicit_d|)$ given that the representatives with respect to S have been calculated. Hence in total step 2 and step 3 take time proportional to the representation of the transition relation δ_d.

Minimizing a guided tree automaton With the splitting operation of the previous section, minimization of a guided tree automaton is now an easy task. Consider a GTA $M_G = (\{Q\}_D, \Sigma, \Delta, \{\bar{q}\}_D, F)$ guided by $G = (D, \mu, d_0)$ and let $N_G = (\{P\}_D, \Sigma, \{\phi\}_D, \{\bar{r}\}_D, H)$ denote the resulting automaton. Now let d be a state space ID with $\mu(d) = (d', d'')$ and assume we have just done a δ_d-split. If the left partition became strictly finer, then we say that a *left-split* occurred. In that case, we must also carry out a $\delta_{d'}$-split operation. Similar considerations apply for a *right-split* and a subsequent $\delta_{d''}$-split operation. This process is repeated until no more split operations need to be done, i.e. until a fixed point has been found.

The algorithm is specified in some more detail in Figure 5.

In the first phase, it performs the split operations according to a set called *candidates*, where ID $d \in candidates$ if a δ_d-split must be carried out. The set *candidates* is updated with respect to the left-splits and right-splits that occur. The first phase of the algorithm terminates when *candidates* = $\emptyset$. The function *split* called with parameter δ performs a split operation with respect to δ as described in the previous section. It returns a pair of Booleans ($lsplit, rsplit$) that indicates whether a left-split (right-split respectively) occurred. The resulting family of partitions, $\mathcal{Q}_D$, is the coarsest, stable family of partitions respecting the initial family of partitions.

In the final phase, the algorithm builds the minimized guided tree automaton from the family of partitions $\mathcal{Q}_D$.

$$\text{Initial family of partitions} : \mathcal{Q}_d = \begin{cases} \{F, Q_{d_0} \setminus F\} & \text{if } d = d_0 \\ \{Q_d\} & \text{otherwise} \end{cases}$$

$candidates = \{d_0\}$
while $candidates \neq \emptyset$ **do**
    remove a state space ID d from $candidates$
    $(d', d'') = \mu(d)$
    $(lsplit, rsplit) = split(\delta_d)$
    if $lsplit$ **then**
        add d' to $candidates$
    fi
    if $rsplit$ **then**
        add d'' to $candidates$
    fi
od

Replace the values in the leaves of the Σ-BDD by canonical representatives according to $\mathcal{Q}_D$ and reduce it. The induced partition of Σ-BDD nodes is denoted S.

for each δ_d of type $d' \times d'' \to d$ **do**
    for each $[q'] \in \mathcal{Q}'$ with canonical representative $(M_{q'}, \hat{\nu}_{q'})$ **do**
        add $[q']$ as a state to P'
        if $\hat{\nu}_{q'} = \perp$ **then**
            $[q'].default_d = \hat{\nu}_{q'}^{\perp}$
            $[q'].explicit_d = M_{q'}^{\perp}$
        else
            $[q'].default_d = \hat{\nu}_{q'}$
            $[q'].explicit_d = M_{q'}$
        fi
    od
    for each $[q''] \in \mathcal{Q}''$ with canonical representative $(M_{q''}, \hat{\nu}_{q''})$ **do**
        add $[q'']$ as a state to P''
        if $\hat{\nu}_{q''} = \perp$ **then**
            $[q''].default_d = \hat{\nu}_{q''}^{\perp}$
            $[q''].explicit_d = M_{q''}^{\perp}$
        else
            $[q''].default_d = \hat{\nu}_{q''}$
            $[q''].explicit_d = M_{q''}$
        fi
    od
od

Figure 5. Algorithm for minimizing GTAs.

Analysis Since each split operation is linear in the total size of the GTA representation and since each operation (except for the last) results in a finer partition, the total running time is $O(n \cdot m)$, where n is the total number of states and m is the total representation size.

Note that the selection of the next transition function to use for a split operation is arbitrary. It would be interesting to study whether more judicious choices could entail asymptotic gains.

It is possible to minimize BDD-represented automata on finite strings in time $O(m \cdot \log m)$ [Kla96], but it is an open question whether this result can be extended to tree automata.

5 Experimental results

The current MONA tool supports *guided tree automata*, but does not yet use the representation of the transitions functions presented in this paper. Instead the implementation uses the BDD encoding of the state spaces mentioned in Section 3. Nevertheless, we have had some successful experimental results with this implementation.

A major goal of the implementation was to provide the means for making FIDO [KS97] a tractable programming language for expressing regular constraints on parse trees. From a FIDO program, a M2L formula is generated. By processing this formula, MONA calculates an automaton, which can be viewed as an attribute grammar for the specified grammar satisfying the syntactic side constraints. The grammar example from [KS97] and an additional HTML grammar example are processed by FIDO and MONA in approximately half a minute on a Sparc Station 1000. In both examples, the M2l formulas generated by FIDO are several (dense) pages long. Our current tool was also used to compute the architectural software constraints in [KKS96].

Our experience is that for most of these examples, the intermediate and final automatons exhibit the property of sparse transition functions. Thus, we expect that the proposed algorithms together with successful attempts of speeding up the current BDD-package will give rise to a significant speed-up in future implementations of the GTA operations.

We have experimented with the guide to determine its practical importance. For the HTML example, we experienced that with a guide with three state spaces, MONA could process the example in 40 seconds, with intermediate automata reaching at most 70 states. With a one-state guide (i.e. with an ordinary DFTA), MONA generates an intermediate automaton with a state space of more than 7000 states—which the subsequent project operation is unable to handle.

References

Bry86.   R.E. Bryant. Graph-based algorithms for boolean function manipulation. *IEEE Transactions on Computers*, C-35(8):677–691, Aug 1986.

CC82.   A. Cardon and M. Crochemore. Partitioning a graph in $O(|A| \log_2 |V|)$. *TCS*, 19:85–98, 1982.

HJJ$^+$96. J.G. Henriksen, J. Jensen, M. Jørgensen, N. Klarlund, B. Paige, T. Rauhe, and A. Sandholm. Mona: Monadic second-order logic in practice. In *Tools* and *Algorithms for the Construction and Analysis of Systems, First International Workshop, TACAS '95, LNCS 1019*, 1996. Also available through http://www.brics.dk/k̃larlund/papers.html.

KKS96.   N. Klarlund, J. Koistinen, and M. Schwartzbach. Formal design constraints. In *Proc. OOPSLA '96*, 1996. to appear.

Kla96.   N Klarlund. An $n \log n$ algorithm for online bdd refinement. Technical report, BRICS Report Series RS-96-, Department of Computer Science, University of Aarhus, 1996.

Koz92.   D. Kozen. On the Myhill-Nerode theorem for trees. *EATCS Bulletin*, 47, 1992.

KS97.   N. Klarlund and M. Schwartzbach. Regularity = logic + recursive data types. Technical report, BRICS, 1997. To appear.

PT87.   R. Paige and R. Tarjan. Three efficient algorithms based on partition refinement. *SIAM Journal of Computing*, 16(6), 1987.

Tho90.   W. Thomas. Automata on infinite objects. In J. van Leeuwen, editor, *Handbook of Theoretical Computer Science*, volume B, pages 133–191. MIT Press/Elsevier, 1990.

Time Series Forecasting by Finite-State Automata

Romuald Boné, Christophe Daguin, Antoine Georgevail
and Denis Maurel

Laboratoire d'Informatique
Ecole d'Ingénieurs en Informatique pour l'Industrie
Université François Rabelais
64 avenue Jean Portalis
37200 Tours, France
E-mail: {bone, maurel}@univ-tours.fr

Abstract. We have developed automata to address the problem of time series forecasting. After turning the time series into sequences of letters, Mohri's algorithm constructs an automaton indexing this text that, once a given word is read, can be used to obtain the set of its positions. By using the automaton to determine what letter usually follows the last sequence of length L, we have developed a one step predictor. This predictor will be used to analyze sunspot activity data.

1 Time Series Prediction

Measurements of temporal fluctuations generate an ordered set of chronological observations (x_t) on the varying value of a statistical series recorded either at successive points in time or over successive periods of time. Such a set is known as a time series. Generating a description of the previously prevalent pattern is a useful first step in projecting the pattern of future data.

The behaviour of a time series reflects a myriad set of influences operating on the variable in question. These influences have traditionally been treated as subject to decomposition into four broad categories:
- Secular trend
- Seasonal variation
- Cyclical fluctuations
- Irregular influences

Certain forces which exert a consistently positive or negative influence on the behaviour of the time series tend to dominate the long-term underlying pattern of the series (eg. changes in technology, population growth). The nature of the influence exerted by such forces is captured by the trend component, which describes the long-term pattern of growth and/or decline in the time series.

Other forces exert either positive or negative influences on the time series, depending on the occasion, in a repetitive, predictable pattern. This recurring pattern, which appears in a time series at regular intervals and which tends to be constant with respect to both timing and amplitude, is referred to as seasonal variation. For example, drink consumption tends to be highest in the summer months, while most suntan lotion is bought in the late spring and summer.

Another type of recurring pattern, brought about by cyclical fluctuations, differs from the seasonal in that the cyclical pattern is not regular with respect to the duration of the pattern, nor to the timing or amplitude from one cycle to the next. Cycle length has tended to vary from 1 to 14 years.

Irregular influences fall into two categories: random influences and nonrecurring influences.

The above four influences are typically considered as interacting to produce observed values of the overall time series:

$$O_t = f(T_t, S_t, C_t, I_t)$$

with

O_t: observed value of time series
T_t: trend component
C_t: cyclical component
S_t: seasonal component
I_t: irregular component.

The international passenger air traffic (thousands of passengers) from January 1949 to December 1960 [1] is shown in Fig. 1. These data present a regular growth (the trend) and an oscillatory effect which is definitely seasonal. Statistical analysis gives a multiplicative model $O_t = T_t * S_t * I_t$.

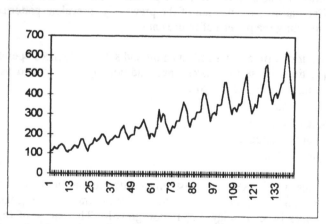

Fig. 1. International passenger air traffic (thousands of passengers) from January 1949 to December 1960

Rather than examining the overall behavior of a time series, it is possible and frequently desirable to decompose the time series into its component elements for separate examination. This process is referred to as decomposition of the time series or time series analysis.

However, this method by itself is often insufficient. A more efficient approach assumes system causality (the present value of x_t is influenced by the past L values of the series) and thus rewrites the relationship as

$$x_t = f\left(x_{t-1}, x_{t-2}, x_{t-3}, \ldots, x_{t-L}\right)$$

where the set $(x_{t-1}, x_{t-2}, x_{t-3}, \ldots, x_{t-L})$ is also termed input-window over time.

The moving average (MA) model, the autoregressive (AR) model ([1]) and the neural network methods ([2], [5], [8]) are each based on this assumption.

2 Indexation by Finite-State Automaton

Finite state automata theory provides a useful tool to index full texts ([4]). The algorithm we employ here, inspired by that of Crochemore [3] and written by Mohri [6], constructs an automaton using alphabet letters for labels, which will give the general index of a text. With this automaton, one can obtain a set of the positions of a given word from the output of the final state (a list of integers).

For instance, the automaton representing the text $t=aabba$ is shown in Fig. 2 11). The output figures are the ending positions of any word reaching this state when read from the initial state. Therefore, to compute the set of positions of a string, one simply has to subtract the length of the string in question from output figures. Since string abb reaches state 4, its position is $4-3=1$.

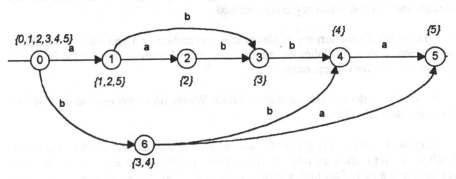

Fig. 2. The automaton representing the text t=aabba

This algorithm is based upon an equivalent relation over factors of text t. Two factors u and v are equivalent if and only if they have the same set of ending positions in t. If they are equivalent, u and v correspond to the same state in the algorithm. In Fig. 2, aab and ab are equivalent.

For building an automaton, the algorithm defines a failure function which allows determination of suffixes of a string u that are not equivalent to u. For instance, aab has a suffixe ab that is equivalent to itself (the set of ending positions in t is {3}) and an other suffixe *b* that is not equivalent to itself (the set of ending positions is {3,4}). Thus the failure function associates state 3 with state 6. Fig. 3 shows this failure function for the automaton of Fig. 2.

The automaton constructed by Mohri is the minimal automaton capable of recognizing the set of suffixes of *t*.

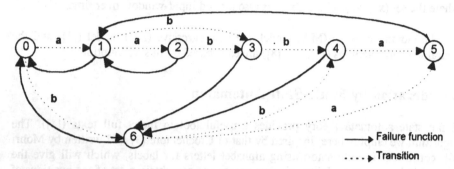

Fig. 3. The failure function of the previous automaton

3 Using the Automaton for Time Series Forecasting

In order to use the automaton, the time series must be converted into a text. First, we remove the trend (if necessary) from the time series. For this purpose, the trend is estimated by any accepted method, including, but not limited to, centered moving average method and linear regression method.

The modified time series y_t is obtained by subtraction (with an additive model) or by division (with a multiplicative model). A statistical study permits us to choose the best parameters for this operation.

Second, we calculate the variations of data. We obtain a new time series $z_t = y_{t+1} - y_t$ with reduced real values.

Then we draw the histogram of z_t and we choose an alphabet for the time series. Each letter represents an interval of variation. In this way we transform the time series z_t into a text. This text is split into two parts: one for automaton construction and another one for the prediction test.

Once this automaton is constructed, we can use it as a one step predictor:
- we choose the length of the input-window over time
- we find with the automaton the letter that usually follows the word that is in the time window
- we predict the value at time t+1, z_{t+1}.

Fig. 4 shows the estimated trend of the international air passenger traffic time series. Since the seasonal component period (determined either by the study of the autocorrelation function or else visually evaluated) is 12 months, the 12-month centered moving average method was used. Fig. 5 shows that the new time series y_t, after removal of the trend. Fig. 6 is the time series $z_t=y_{t+1}-y_t$ (variations of the time series y_t). Fig. 7 represents the histogram of the time series z_t.

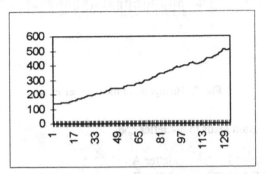

Fig. 4. Estimated trend of the time series

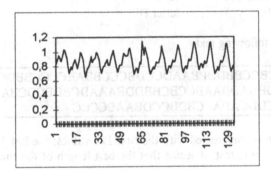

Fig. 5. Removal of the trend

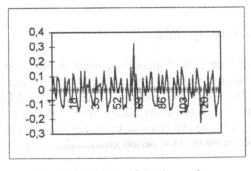

Fig. 6. Variations of the time series y_t

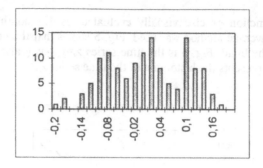

Fig. 7. Histogram of the time series z_t

Studying the histogram leads to a 4 letter code:

$z_t < -0.1$ letter A
$-0.1 \leq z_t < 0$ letter B
$0 \leq z_t < 0.1$ letter C
$0.1 \leq z_t$ letter D

which creates the following text:

CCBBCCBBABCBCCBBDDBBAADCCDBCCCBBABCCCCBBDCCABBCCBDB
BCCCABACCBDBDADBAABCCBCBBDDBAAADCBDBCDCBAABCCBDBCD
CBAABCCBDBCDDCABACCBDBCCDBAABCCCC

The automaton is constructed with the first 120 values; the last 14 values will be used for the forecasting test. It seems that the best length of the time window is 11 letters.

The success rate of the letter prediction is about 65% on the 14 values, an interesting result by comparison with a random walk (25 %).

4 The Sunspots Activity Data

The Sunspot series (Priesley, 1988) gives the annual activity of sunspots since 1700 (Fig. 8). These nonstationary series are used classically to compare and evaluate statistical modeling and forecasting methods. These series are fairly typical of the kind of material with which our theory deals: data that presents a very irregular fluctuation with, so far as the eye can see, no systematic element, and no tendency towards increase or decrease over the period given.

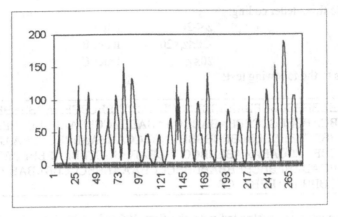

Fig. 8. The Sunspot series

Fig. 9 shows variations of the time series, while Fig. 10 represents the histogram of the time series.

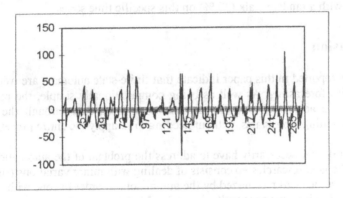

Fig. 9. Variations of the times series

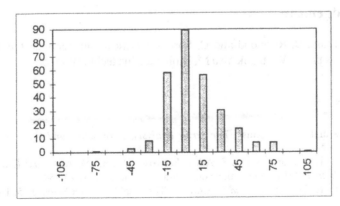

Fig. 10. Histogram of the time series

We choose a three letter coding:

$z_t < -20$ letter A
$-20 \leq z_t < 20$ letter B
$20 \leq z_t$ letter C

which leads to the following text:

```
BBBBCABBBBBBBBBBCBBABBBBBBCCBAABABBBCBCBAABBBBBBCCBAB
BBBBCBBBCABBBBCCCBBBABABCCAABABBBCBBAABBBBBBBBBBBBBB
BBBBBBBBBBCBBBBBBBBBBBBBBBAABBCCACBAABBBCCCCAABBBBBB
BCCBBBBBBBBCCCABAAABBBBCCBBBBABBBBCCBBBABBBBBBBBCBBB
BABBBBCBCABABBBBCBBBBABBBBCCCBAAABBBCCCBBABABBCCCBA
AABBBBCCBBBABABBBBCCB
```

The automaton is constructed with the first 265 values, with the last 15 values reserved for a forecasting test. A length of 12 letters was chosen for the time window.

The success rate of the prediction is about 67%, an interesting result by comparison with a random walk (33 %) on this specific time series.

5 Conclusion

The analyses reported in this paper indicate that finite-state automata are well suited for time-series forecasting. As well as being computationally simple, the presented method has the advantage of easily processing updated data through the simple procedure of adding more states, without needing to re-apply the entire process.

Future work will necessarily have to address the problem of the automatic choice of coding. The next research step consists of dealing with minor variations (such as a single letter) in the word recognized by the automaton, in order to cope with noise in time series and to improve our results.

Acknowledgements

The authors thank D. Raymond and C. Sori for critical comments on the different versions of this paper. We thank also M. Mohri for his technical report.

References

1. G.E.P. Box and G.M. Jenkins. *Time series analysis, Forecasting and Control*, Holden-Day, San Francisco, 1970.
2. J.T. Connor, R.D. Martin and L.E. Atlas. *Recurrent Neural Networks and Robust Time Series Prediction*, IEEE Transactions on Neural Networks, 5(2):240-254
3. M. Crochemore. *Transducers and Repetitions*, Theorical Computer Science 45, 1986.

4. M. Crochemore and W. Rytter. *Text Algorithms*, Oxford University Press, 1994.
5. A. Kouam, F. Badran F., S. Thiria.. *Approche Méthodologique pour l'Etude de la Prévision à l'aide des Réseaux de Neurones*, Cinquièmes Journées Internationales, 2-6 Novembre 1992, Nîmes, France.
6. M. Mohri. *Application of Local Grammars Automata: an efficient Algorithm*, Technical repport IGM 94-16, University of Marne-la-Vallée, France,1994.
7. M.B. Priestley. *Non-linear and Non-stationary Time Series Analysis*, Academic Press, 1988.
8. Weigend A. S., Gershenfeld N.A. (eds.). *Time series Prediction, Forecasting the Future and Understanding the Past*, Proceedings of the NATO Advanced Research Workshop on Comparative Time Series Analysis, Santa Fe, New Mexico, 1992.

Dynamical Implementation of Nondeterministic Automata and Concurrent Systems

Max Garzon[1] and Eugene Eberbach[2] *

[1] Institute for Intelligent Systems
[2] Department of Mathematical Sciences, The University of Memphis,
Memphis, TN 38152-6429,

Abstract. Several recent results have implemented a number of deterministic automata (finite-state, pushdown, even Turing machines and neural nets) using piecewise linear dynamical systems in one- and two-dimensional euclidean spaces. Nondeterministic devices have only been likewise implemented by iterated systems containing several maps. We show how to simulate nondeterministic and concurrent systems (finite-state automata, pushdown automata, and Petri nets) using a single deterministic piecewise linear map of the real interval. As a consequence, we establish a correspondence between nondeterminism and incremental entropy of the corresponding dynamical system. Relationship to the separation of complexity classes is discussed.

1 Introduction

It is now a well established fact that dynamical maps of the continuum (e.g., the real interval) that admit a very simple specification can have very complex dynamical behavior, as witnessed by the presence of properties such as chaos, sensitivity to initial conditions, etc. [3]. A more computational approach has subsequently been taken to investigate their complexity using tools from the theory of sequential computation. Particularly well known are, for example, the so-called kneading sequences that map orbits to one-way infinite words and dynamical systems to formal languages [5]. More recently, the complexity of piecewise linear maps in euclidean spaces has been investigated in a dual direction, namely, in terms of the computational devices they implicitly carry within. For example, it has been shown that they are complex enough in dimension two to implement arbitrary computing procedures in the form of Turing machines, although this is impossible on the unit interval [2]. This question can be precisely interpreted as asking to what extent man-made computational models such as finite-state machines, Turing machines, artificial neural nets and the like can be implemented *in nature*, as it is modeled by dynamical systems on the continuum.

It has been further shown that piecewise linear maps of the interval can nonetheless implement in real-time deterministic finite-state machines and even

---

* This work was done while the second author was on sabbatical leave from the Jodrey School of Computer Science, Acadia University, Wolfville, Nova Scotia, Canada B0P 1X0 at The University of Memphis.

implement the combinatorial decision making embodied in deterministic push-down automata recognizing arbitrary linear context-free languages under robust continuous recursive two-to-one encodings [1]. The same paper further shows that by relaxing the first two conditions (robustness and continuity) and requiring only isomorphism of an attractor subbasin in the simulation, one can implement arbitrary algorithms, as specified by deterministic halting Turing machines, on piecewise linear maps of the real interval.

Thus, the fundamental difficulty in addressing this question lies in the precise meaning of the notion of implementation. The dynamical systems point of view is a most natural approach. Both maps on the continuum and computational models, although acting on different domains, can be regarded as dynamical systems on a metric space. An implementation thus requires an encoding of the total states of the computational device (e.g., a finite-state machine or a Petri net) in states in the continuum, e.g., real numbers. Ideally, the encoding should be bijective and bicontinuous, i.e., a full conjugacy between the two systems, but it is obvious that the state space of a computational device, being discrete and perhaps countable, cannot be homeomorphic to a continuum. Hence, one must relax the constraints on the encodings, and rather insist on properties such as injectivity, recursivity, etc.

So far, implementations have been restricted to deterministic automata due to the obvious constraint that a dynamical system is functional and hence deterministic. It is relatively easy to show that nondeterministic pushdown automata can be implemented by a nondeterministic set of piecewise linear maps on the real unit interval I [7,8]. In this paper we explore the natural question whether it is possible at all to implement *nondeterministic* automata using a *single deterministic dynamical system* on the interval. The interesting issue about nondeterminism and concurrency, however, is to what extent they are more powerful computing paradigms than the corresponding deterministic systems, which is the problematic and key component of many current and old-standing computational questions. It would be of interest to know what relationship exists, if any, between this problem and analogous problems about dynamical systems on the interval. We present a preliminary solution to this question. Dynamical systems on continuous spaces, particularly euclidean spaces, have been studied for many years and some of their behavior is better understood than that of abstract automata. The results of this paper show that indeed, it is possible to implement nondeterministic automata and concurrent systems using deterministic dynamical systems, and, moreover, that nondeterminism is closely related to the important and sometimes characteristic notion of entropy in dynamical systems.

The rest of this paper is organized as follows. Section 2 defines precisely the notion of implementation of an automaton by a dynamical system and shows how to implement arbitrary finite-state automata. Section 3 briefly introduces Petri nets and shows an implementation by dynamical systems. Section 4 describes pushdown automata as dynamical systems and shows how to implement them likewise. Finally, Section 5 presents some preliminary observations about the resulting relationship between nondeterminism and entropy.

2 Implementation of Nondeterministic Finite-state Automata

A *dynamical system* on a metric space X is simply a continuous self map T : $X \to X$. Let I denote the unit interval $[0, 1]$. In keeping with the notation in [2], PL_d denotes the set of piecewise-linear continuous functions on I^d. More precisely, $f : I^d \to I^d$ belongs to PL_d if

- f is continuous,
- there is a sequence $(P_i)_{1 \le i \le p}$ of convex closed polyhedra (of nonempty interior) such that $f_i = f_{|P_i}$ is affine, $I^d = \bigcup_{i=1}^{p} P_i$ and $\overset{\circ}{P_i} \cap \overset{\circ}{P_j} = \emptyset$ for $i \ne j$,

where $stackrel \circ P_i$ denotes the interior points of a set P (i.e., those that can centered in a ball of positive radius tucked inside P). In the case $d = 1$, a convex polyhedron P_i is just an interval $I_i = [c_i, c_{i+1}]$, on which the corresponding affine map is of the form $f_i(x) = a_i x + b_i$ for $x \in [c_i, c_{i+1}]$. The reader is referred to Devaney [3] for background on dynamical systems.

Definition 1. Let T be the transition function of a machine M (which may be a a finite-state machine, a pushdown automaton, a Turing machine, etc.) and let C be its configuration space. A function $f : I^d \to I^d$ implements (or simulates) M if there exists an encoding $\Phi : C \to I$ such that

$$T = \Phi^{-1} \circ f \circ \Phi.$$

Intuitively, this means that in order to simulate one step of M, one can encode its configuration (i.e., the current state and tape contents) with Φ, apply f, and then get the result by decoding with Φ^{-1}.

A system T is said to implement another system M in real time (respectively, linear time, quadratic time, etc.) if a computation of M for t time units can be performed by T in time t (respectively, $O(t), O(t^2), etc.$).

The encoding mapping $\Phi : C \to I$ should be continuous and injective. It is sufficient to consider the standard Cantor set encoding, i.e., the radix-3 expansion of C (real numbers not containing the digit 1) over the binary alphabet $\{0, 2\}$, and which is homeomorphic (i.e., topologically isomorphic) with the infinite Cartesian product $\{0, 2\} \times \{0, 2\} \times \cdots$, i.e., infinite strings over the alphabet $\{0, 2\}$. Another solution is to apply the Cantor-like encoding, used already in [10,2], with the radix-4 expansion on the binary alphabet $\{1, 3\}$ (i.e., not containing digits 0 and 2, thus, in some sense "complementary" to the Cantor encoding), and which is homeomorphic with the Cantor set.

Definition 2. A finite-state automaton, or finite-state machine (fsm for short), M is a 5-tuple $M = \langle Q, \Sigma, \delta, q_0, F \rangle$ consisting of an alphabet Q of memory configurations (i.e., finite states) with special symbol q_0 (initial state), an alphabet

Σ of input symbols containing a "blank symbol" b and a "reset symbol" r, a transition function (called dynamics)

$$\delta : Q \times \Sigma \to 2^Q$$
$$\delta(q, x) = \{q_1, ..., q_n\} \,.$$

For a deterministic fsm, $\delta : Q \times \Sigma \to Q$ is really single-valued $\delta(p, x) = q$, but for a nondeterministic machine, δ may be multi-valued. Alternatively, for a nondeterministic fsm, we use throughout the paper the determinized form with an auxiliary index set Y labeling the transitions, given by

$$\delta : Q \times \Sigma \times Y \to Q$$

$$\delta(q, x, y_i) = q_i \ (1 \leq i \leq n) \,.$$

Finally, $F \subseteq Q$ is a distinguished subset of final states. The blank symbol b denotes the end of input (no more input symbols), and the reset symbol r "resets" the fsm to its initial state q_0.

A finite-state machine is intended to be a mathematical model of a primitive sequential computer that can read but not write on or change its input. The reader is referred to [4] for further background on automata theory. We interpret a fsm as a dynamical system $T : \mathbf{C} \to \mathbf{C}$. A *configuration* of M is an element of $\mathbf{C}_0 := Q \times \Sigma^*$ describing the total state of M's computation (internal state plus input yet to be read) so that we can define recursively (identifying a configuration (q, x) with its encoding $\Phi(q, x)$),

$$T(q, x_1 x_2 ... x_n) := (\delta(q, x_1), x_2 ... x_n) \,.$$

Likewise, a configuration of a nondeterministic fsm is an element of $\mathbf{C}_0 = Q \times (\Sigma \times Y)^*$, so that

$$T(q, x_1 y_1 x_2 y_2 ... x_n y_n) = (\delta(q, x_1, y_1), x_2 y_2 ... x_n y_n) \,.$$

The reader is referred to [6] for further background on automata as dynamical systems.

Now we are ready to show how to implement nondeterministic fsm by deterministic piecewise linear maps of the interval.

Theorem 3. *An arbitrary nondeterministic finite-state automaton can be implemented in linear time by a piecewise linear map in* PL_1.

Proof. We implement nondeterministic automata by simultaneously simulating all possible nondeterministic runs in one dynamical update. The state of an automaton, an input sequence (including the blank and reset symbols) and indices defining the choices made in a run, are encoded over the alphabet $\{0, 2\}$ in the radix-3 expansion of a point $x \in I$, i.e., it is a standard Cantor set encoding. We assume that the cardinalities $|Q| = |Y| \leq 2^p$ and $|X| \leq 2^m$, i.e. that each state symbol and each index will require at most p binary digits, and each input symbol at most m binary digits. Therefore the encoding of the total state (q, x) of the automaton is

$$x = 0.qx_1y_1^1...x_ny_n^1 \, r \, y_r x_1y_1^2...x_ny_n^2 \, r \, y_r \cdots x_1y_1^k...x_ny_n^k by_b by_b by_b \cdots$$

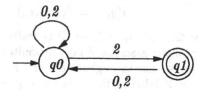

Fig. 1. A simple nondeterministic fsm

Let $q' := \delta(q, x_1, y_1^1)$ be the next state of the fsm (determined by the current state q, the input character x_1 and the index symbol y_1^1 identifying the nondeterministic choice being made). The function f is defined as follows.

- $f(x) := 0.q' + (x - 0.qx_1y_1^1) * 3^{p+m}$ for pairs: an "ordinary" input symbol with its index symbol;
- $f(x) := 0.q_0 + (x - 0.qry_r) * 3^{p+m}$ for the reset symbol r with its index y_r (it always causes f to return to the initial state q_0)
- $f(x) := x$ for the blank symbol b with its index symbol y_b (no movement).

All three equations are clearly linear and give rise to a piecewise linear map with mp^2 linear pieces (p values for q and y_1^1, and m values for x_1). Because there are gaps in the Cantor encoding, all pieces of the "valid" encoding can be connected to get a continuous piecewise linear map.

We will illustrate with a simple nondeterministic fsm.
Example 1. Let the automaton in Fig. 1 have only two states q_0 and q_1, where q_0 is the initial state, q_1 is the terminal state, the input alphabet $\{0, 2\}$, and the transition function (written in two equivalent forms (without and with index symbols).

It is easy to show that the nondeterministic fsm of Fig. 1 recognizes the language $L = (0 \cup 2)^*2(0 \cup 2)^*$. The determinized transition function is

$\delta(q_0, 0) = q_0$ or $\delta(q_0, 0, 0) = \delta(q_0, 0, 2) = q_0$

$\delta(q_0, 2) = \{q_0, q_1\}$ or $\delta(q_0, 2, 0) = q_0, \delta(q_0, 2, 2) = q_1$

$\delta(q_1, 0) = q_0$ or $\delta(q_1, 0, 0) = \delta(q_1, 0, 2) = q_0$

$\delta(q_1, 2) = q_0$ or $\delta(q_1, 2, 0) = \delta(q_1, 2, 2) = q_0$

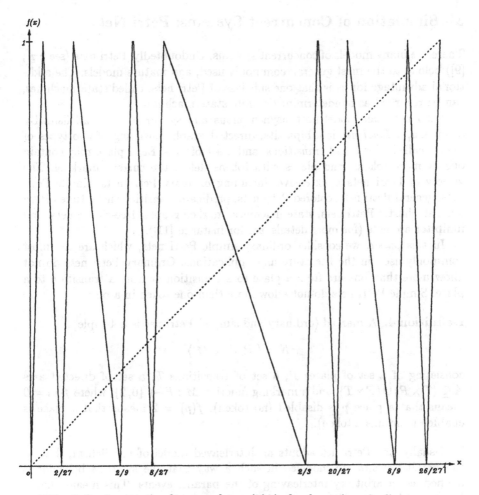

Fig. 2. Implementation by a nondeterministic fsm by a piecewise-linear map

For notational simplicity, we will illustrate only one run in the encoding, and we will omit the blank symbols. We can encode nondeterministic runs in the standard Cantor as the point $x \in I$ given by

$$x := 0.qx_1y_1x_2y_2...x_ny_n .$$

The linear piece implementing the elementary move of the fsm $\delta(q, x_1, y_1) := q'$, will be

$$f(x) := 0.q' + (x - 0.qx_1y_1) * 3^2 .$$

The dynamical map consists of 8 linear pieces used actively for simulation, and of seven linear pieces added to have a continuity. It is presented in Fig. 2.

Note that adding blank and reset symbols gives a map with 16 active pieces and 15 pieces connecting them for continuity.

3 Simulation of Concurrent Systems: Petri Nets

There are many models of concurrent systems. Undoubtedly, Petri nets (see e.g., [9]) belong to the most general, commonly used, and mature models. The additional advantage for us is that one subclass of Petri nets, called state machines, can be regarded as nondeterministic finite-state machines.

Petri nets can model both asynchronous and synchronous parallelism. By definition, a *Petri net* is a bipartite, directed graph consisting of two types of nodes, called *places* and *transitions*, and a set of *arcs*. Each place may contain one or more tokens, and places with tokens define the current marking. The variety of Petri nets is extensive, including ordinary Petri nets, simple Petri nets, general Petri nets, colored Petri nets, predicate transitions nets, timed Petri nets, stochastic Petri nets, state machines, marked graphs, free-choice nets, and inhibitor-arc nets (for more details, see for instance [11]).

In this paper, we consider ordinary simple Petri nets, which are the most commonly used in the literature and applications. Ordinary Petri nets do not allow more than one arc from a place to a transition or from a transition to a place. Simple Petri nets do not allow more than one token in a place.

Definition 4. A marked (ordinary and simple) Petri net is a 4-tuple

$$PN = \langle\, P, T, A, M\, \rangle$$

consisting of a set of places P, a set of transitions T, a set of directed arcs $A \subseteq (T \times P) \cup (P \times T)$, and a marking function $M : P \to \{0, 2\}$, where $f(p) = 0$ means that a place p is disabled (no token), $f(p) = 2$ means that a place is enabled (contains a token).

Usually, the Petri net adopts an interleaved model of parallelism, that is, the parallel events are modeled in such a way that the system's behavior is defined as an arbitrary interleaving of the parallel events. This means that if two transitions t_1 and t_2 are enabled in some marking and are not in conflict (i.e., they are independent in the sense that the intersection of the input/output places is empty), then they can fire either in the given or in the opposite order. Such transitions represent events that can occur in parallel. The whole input to a Petri net (working as an interleaved sequence recognizer) can then be considered as concatenation of all possible interleaved sequences of transitions.

For the purpose of this paper, we assume that the set of places is ordered, i.e., $P = (p_1, ..., p_m)$. The current state (marking) of the Petri net is encoded as $Q = (M(p_1)...M(p_m))$. For example, the Petri net state with 4 places, where the first and the third place are enabled, can be encoded as (2020). We can now define the transition function of the Petri net as a function

$$\delta : Q \times T \times Y \to Q$$
$$\delta(q, t, y) := q',$$

where the index set Y allows to select a transition in a unique way (determined both by $|P|$ and the number of nondeterministic choices). We interpret a Petri net as a dynamical system $T : \mathbf{C}_0 \to \mathbf{C}_0$ as for finite-state machines. Thus $\mathbf{C}_0 := Q \times (T \times Y)^*$ are also configurations of a (possibly nondeterministic) Petri net, and, identifying a total state of the net with its encoding as before, $T(q, t_1 y_1 x_2 y_2 ... x_n y_n) = (\delta(q, t_1, y_1), t_2 y_2 ... t_n y_n)$.

Theorem 5. *An arbitrary (simple and ordinary) (possibly nondeterministic) Petri net can be implemented in a linear time by a piecewise linear map in PL_1.*

Proof. The proof is essentially the same as for nondeterministic fsm. The simulation will be generally the same, whether we will assume that the Petri net is deterministic or nondeterministic. Nondeterminism will affect only the size of the index set Y and thus the number of linear pieces used in the implementation. The number of configurations (runs) will be affected both by nondeterminism and the number of interleaved transition sequences

As an example we will consider a piecewise-linear simulation of the Petri net representing one of the standard problems from concurrency theory, namely the producer-consumer problem.

Example 2. In this problem, a process produces items that are consumed by a consumer process using an intermediate buffer. The rate at which the the items are produced is independent of the rate at which they are consumed, but it is limited (synchronized) by the capacity of a buffer (equal to 1 in this example).

The initial and terminal states are the same (200200), i.e. p_1 and p_4 are enabled. The "waste-basket" place p_6 has been added to allow all possible transitions to be fired for each possible marking (to allow all possible sequences of transitions). Examples of two possible initial transitions are $\delta((200200), p, 1) = (020200)$ and $\delta((200200), r, 3) = (200020)$.

We can encode all interleaved sequences by the real number

$$x = 0.q t_1^1 y_1^1 ... t_n^1 y_n^1 r y_r t_1^2 y_1^2 ... t_n^2 y_n^2 r y_r ... t_1^k y_1^k ... t_n^k y_n^k b y_b b y_b b y_b ...$$

where k will be larger if more interleaving/parallelism is available.

To simplify the problem we only show one interleaved run, without using the blank and reset symbols.

$$x = 0.q t_1 y_1 ... t_n y_n$$

The simulation map will be the following

$$x' = f(x) = q' + (x - 0.q t_1 y_1) * 3^4$$

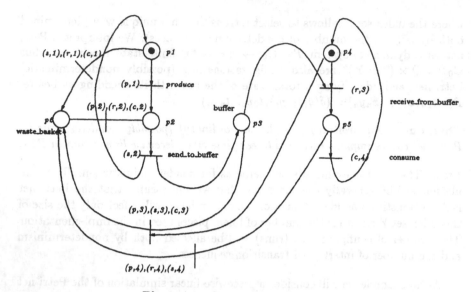

Fig. 3. A producer-consumer model.

4 Simulation of Nondeterministic Pushdown Automata

The ideas of the implementation of deterministic finite-state machines by piecewise linear maps have been extended to deterministic pushdown automata in [1]. Based on this work, it is straightforward to extend prior results to nondeterministic pushdown automata via analogous encodings. We summarize these results in this section.

A pushdown automaton M is interpreted as a dynamical system $T : \mathbf{C} \to \mathbf{C}$, where $\mathbf{C}$ is the set of instantaneous descriptions of M endowed with the product topology. We use essentially the same notion of implementation of a machine by a dynamical system of the previous section. Let $M = \langle \Sigma, Q, \Gamma, \delta \rangle$ be a deterministic pushdown automaton, where Σ denotes a finite input alphabet, Q denotes a finite state set, Γ denotes a finite stack alphabet, and

$$\delta : \Sigma \times Q \times \Gamma \to Q \times \Gamma$$

is the transition function, (see [4] for more details). An instantaneous description of M is a triple (x, q, y), where $x \in \Sigma^*$, $q \in Q$ and $y \in \Gamma^*$. We identify each such triple with a finite string and a biinfinite sequence as follows.

$$(x, q, y) \equiv \quad \iota \ x_n \ \cdots \ x_1 \ q \ y_1 \cdots y_m \ \iota$$
$$\equiv \cdots \iota \ \iota \ z_{-n} \ \cdots z_{-1} \ z_0 \ z_1 \cdots z_m \ \iota \ \iota \ \cdots$$

where $x = x_1 \cdots x_n \in \Sigma^*$ denotes the input string, x_1 denotes the current input, $y = y_1 \cdots y_m \in \Gamma^*$ denotes the current stack configuration (y_1 is the symbol at the top of the stack). We denote by λ, the blank or null character string, and by

ι, the end-of-input and end-of-stack character. Let $S := \Sigma \cup Q \cup \Gamma$. We may identify $\mathbf{C}$ as a dense subset of a closed and open subset of $S^{\mathbf{Z}}$, the biinfinite sequences over S, i.e.,

$$\mathbf{C} \subset S^{\mathbf{Z}} = \{z = \cdots z_{-1} z_0 z_1 \cdots \mid z_i \in S\}$$

where Z denotes the set of integers $\{\ldots, -2, -1, 0, 1, 2, \ldots\}$. We next define a (so-called astronomer's) metric $|*, *| : S^{\mathbf{Z}} \times S^{\mathbf{Z}} \to [0, 1]$ that gives rise to the product topology on $S^{\mathbf{Z}}$. There are several choices, one of which is the following.

$$|z^{(1)}, z^{(2)}| := |S|^{-k} \quad k := \min\left\{|i| : z_i^{(1)} \neq z_i^{(2)}\right\}$$

The *cylinder set* defined by a finite string ω consists of all configurations that contain ω as a substring centered at the origin. Note that we have defined $\mathbf{C}$ in such a way that the action of the transition function $\delta(x, q, y) \mapsto (x', q', y')$ corresponds to an action on cylinder sets

$$[z_{-1} z_0 z_1] = [x\ q\ y] \to [x'\ q'\ y'] = [z'_{-1} z'_0 z'_1]$$

Therefore T is a continuous map on $\mathbf{C}$, and we can still regard M as a dynamical system $T : S^{\mathbf{Z}} \to S^{\mathbf{Z}}$ given by

$$T(z \equiv (x, q, y)) := z' \equiv (x', q', y') \text{ iff } \delta(x, q, y) \to (x', q', y')$$

and $T(z) := z$ otherwise. Note that $T : S^{\mathbf{Z}} \to S^{\mathbf{Z}}$ does not change its argument where $z_k = q$, for $k \neq 0$, i.e., where a state symbol appears in either the input string x or in the contents of the stack y, or where $z_0 \in \Sigma \cup \Gamma$, i.e., where an input or stack symbol appears at z_0. We are therefore taking the convention that M immediately halts and rejects whenever an undefined state, input or stack symbol is encountered. Also note that there may be nonhalting configurations $z \in S^{\mathbf{Z}}$, where $z_i \neq \iota$ for all i, i.e., where either the input has no end-of-input marker, or the stack has no bottom-of-stack marker.

The $K = |S|$ cylinder sets $[s_k]$, $s_k \in S$, form a natural cover of $\mathbf{C}$, and it would be desirable to have $\Phi([s_k]) \subseteq [a_k, b_k]$, for $k = 1, \ldots, K$, where the (perhaps overlapping) intervals $[a_k, b_k]$ form a cover of $[0, 1]$. If we can in fact extend Φ to $S^{\mathbf{Z}}$ so that $\Phi([x_1\ q\ y_1]) = J_{x_1, q, y_1}$ is an interval in $[0, 1]$, then the 'transitions' $T([x\ q\ y]) = [x'\ q'\ y']$ will correspond to 'interval exchanges' $f(J_{x,q,y}) = J_{x',q',y'}$. Indeed, the cylinder sets are closed subsets, and hence themselves isomorphic to ternary Cantor sets; we can thus close the gaps of the Cantor set up onto an interval so as to obtain a semiconjugacy (except for 0 and 1 in the Cantor set) to produce a recursive encoding. (Such a map sending covers to covers is called a *robust encoding* in [1]. A slightly different version is defined in [2] as *interval encodings*.) From a different view, one may regard the simulation of M by f as being *robust*, in the sense that there exist $\epsilon > 0$ such that any point $p \in (\Phi(z) - \epsilon, \Phi(z) + \epsilon)$, for $z = (q, x, y) \in \mathbf{C}$, will have the same itinerary i.e., visit the same sequence of partition subsets as $\Phi(z)$, and hence can be initially used to encode z just as well.

With this encoding and a similar proof we thus obtain

Theorem 6. *Every nondeterministic pushdown automaton can be implemented by a piecewise linear map of the interval.*

5 Nondeterminism and Entropy

In this section we show an application of this type of implementation using entropy. The notion of entropy has been used in several contexts. Entropy is a classical notion in physics, where it was introduced as a measure of disorder, particularly in thermodynamics. For dynamical systems, the notion of entropy quantifies the dynamical complexity of the iteration of a dynamical system. There are at least two types of entropy for dynamical systems: metric and topological. We will use the latter since it is more general and does not require background in measure theory [6].

An orbit of a dynamical system f is a sequence $\{x^t\}_{t\geq 0}$ such that $f(x^t) = x^{t+1}$ for $t \geq 0$. A basic problem in the study of dynamical systems is to identify the long-term behavior of its orbits.

Definition 7. Let (X, d) be a compact metric space and $T : X \to X$ be a continuous self-map of X. Let n be a nonnegative integer and $\varepsilon > 0$. A subset $E \subseteq X$ is called (n, ε)-spanning if for every $x \in X$ there exists $a \in E$ such that $|T^t(x), T^t(a)| < \varepsilon$ for $t = 0, 1, \cdots, n-1$.

The smallest cardinality of an (n, ε)-spanning set is here denoted $n_{T,\varepsilon}$, or just n_ε if T is understood. Since X is compact, n_ε is finite. It usually grows exponentially fast with n. The growth rate of n_ε is measured by $\limsup \frac{1}{n} \log n_\varepsilon$.

Definition 8. The topological entropy of T is given by

$$ent(f) = \lim_{\varepsilon \to 0+} \limsup \frac{1}{n} \log n_\varepsilon.$$

Intuitively speaking, a spanning set gives a complete sample of starting points whose orbits approximate, to within a degree of tolerance specified by a distance ε, initial segments of arbitrary orbits for a given length of time n. Entropy can be understood as a logarithmic measure of how fast the orbits of a dynamical system are collapsed together for a given observational precision size (determined by ε) and the number of observed configurations (defined by n). The entropy of a dynamical system is a real number that varies continuously from 0 to, and possibly including, ∞. Moreover, the entropy is invariant under conjugacies, i.e., two dynamical systems equivalent under a homeomorphic change of variables (such as equivalent metrics) have the same entropy.

Under the correspondence with dynamical systems established by the previous sections, one can obtain the following results relating nondeterminism and entropy.

Theorem 9. *The entropy of a dynamical system that implements an automaton by an injective encoding is an invariant of a deterministic automaton. More precisely, if Σ is the input alphabet, $|Q|$ the state set, and Γ the stack alphabet, then*

1. *The entropy of a finite-state automaton is either exactly $\log|\Sigma|$ (if it can use only a bounded finite number of nondeterministic choices), or at most $\log|\Sigma|+\log|Q|$ (if it makes an unbounded number of nondeterministic choices).*
2. *The entropy of a pushdown automaton is either exactly $\log|\Sigma| + \log|\Gamma|$ (if it can use only a bounded finite number of nondeterministic choices), or at most $\log|\Sigma| + \log|Q| + \log|\Gamma|$ (if it makes an unbounded number of nondeterministic choices).*

In particular, the entropy increases with increasing nondeterminism (although perhaps not smoothly).

Proof. For simplicity, for astronomer's metric we use the equivalent metric given for two configurations x and a by

$$|x,a| := \Sigma_i \frac{|x_i - a_i|}{3^{|i|}} \, .$$

In the astronomer's metric, a ball of radius $\varepsilon > 0$ centered at a configuration a amounts to a so-called cylinder set consisting of all those configurations that agree with a inside an ε-window of radius p, which is roughly equal to the number of significant ternary digits in $1/\varepsilon$. To find the minimal cardinality of an (n,ε)-spanning set for a given $\varepsilon > 0$, finitely many configurations a^k have to be chosen in order that every finite initial segment of an orbit $\{T^t(x)\}_t$ be traceable within ε by the orbit of some a^k, i.e., $|T^t(x), T^t(a)| < \varepsilon$. If p symbols fall inside the ε-window, the maximal error caused by neglecting the symbols outside the window for a fsm will be

$$\frac{2}{3^{p+1}} + \frac{2}{3^{p+2}} + \cdots = \frac{1}{3^p} < \varepsilon.$$

This gives that the size of the window should be $p > -c\log\varepsilon$, where c is a constant. The same result (with a different value for the constant c) will be valid for pda. If there were no moves ($n = 0$), the cardinality of a minimum spanning set E thus clearly equals $2^{-c\log\varepsilon}$ since each a_i inside the ε-window may take two possible values. However, for larger n, one must include appropriate values in a larger window for the a^k's so that additional differences that may shift into the window in the first $n - 1$ simulation moves may be traced. This gives

$$n_\varepsilon = 2^{p+\#stack+\#input+\#index} ,$$

where $\#_{stack}$ ($\#_{input}$, $\#_{index}$) is the number of stack (input, index, respectively) binary digits. Therefore

$$
\begin{aligned}
ent(f) &= \lim_{\epsilon \to 0^+} \limsup \frac{1}{n} \log n_\epsilon \\
&= \lim_{\epsilon \to 0^+} \limsup \frac{p + \#_{stack} + \#_{input} + \#_{index}}{n} \log 2 \\
&= \lim_{\epsilon \to 0^+} \limsup \frac{\#_{stack} \#_{input} + \#_{index}}{n} \log 2.
\end{aligned}
$$

Now it is clear that to find n_ϵ, it is sufficient to consider only the maximum number of values taken by symbols falling inside the window during $n - 1$ simulation steps. We need to consider separately fsa and pda with bounded and unbounded nondeterminism.

1. A fsm with a bounded finite number of nondeterministic choices:
   In this case, $\#_{stack} = 0$, $\#_{input} = (n - 1) \log_2 |\Sigma|$, $\#_{index} = k$, where k is a constant independent of n. One can interpret a finite number of nondeterministic choices as a finite number of index values. In particular, a deterministic automaton can be interpreted as having all index values fixed. Thus

$$
ent(f) = \lim_{\epsilon \to 0^+} \limsup \frac{(n - 1) \log_2 |\Sigma| + k}{n} \log 2 = \log |\Sigma|.
$$

2. A fsm with an unbounded number of nondeterministic choices:
   In this case, $\#_{stack} = 0$, $\#_{input} = (n - 1) \log_2 |\Sigma|$, $\#_{index} \leq (n - 1) \log_2 |Q|$ since in every step we have to consider in the worst case $|Q|$ possible nondeterministic choices, represented by $\log_2 |Q|$ binary digits in the Cantor encoding. Therefore,

$$
ent(f) \leq \lim_{\epsilon \to 0^+} \limsup \frac{(n - 1) \log_2 |\Sigma| + (n - 1) \log_2 |Q|}{n} \log 2 = \log |\Sigma| + \log |Q|.
$$

3. A pda with a bounded finite number of nondeterministic choices:
   In this case, $\#_{stack} = (n - 1) \log_2 |\Gamma|$, $\#_{input} = (n - 1) \log_2 |\Sigma|$, $\#_{index} = k$. Therefore,

$$
ent(f) = \lim_{\epsilon \to 0^+} \limsup \frac{(n - 1) \log_2 |\Sigma| + k + (n - 1) \log_2 |\Gamma|}{n} \log 2 = \log |\Sigma| + \log |\Gamma|.
$$

4. A pda with an unbounded number of nondeterministic choices:
   In this case, $\#_{stack} = \log_2 |\Gamma|$, $\#_{input} = (n - 1) \log_2 |\Sigma|$ and $\#_{index} \leq (n - 1) \log_2 |Q|$. Therefore,

$$
\begin{aligned}
ent(f) &\leq \lim_{\epsilon \to 0^+} \limsup \frac{(n - 1) \log_2 |\Sigma| + (n - 1) \log_2 |Q| + (n - 1) \log_2 |\Gamma|}{n} \log 2 \\
&= \log |\Sigma| + \log |Q| + \log |\Gamma|. \quad \square
\end{aligned}
$$

6 Conclusions and Open Problems

This paper shows how relatively simple deterministic dynamical systems in the continuum are capable of implementing nondeterministic and concurrent computation. Specifically, piecewise linear maps of the interval can be used to implement finite-state machines and pushdown automata, both deterministic and nondeterministic, as well as concurrent systems such as Petri nets. These implementations make digital computer implementations readily programmable in any numerical package, such as Maple or Mathematica, and surely faster and more efficient in floating point in high speed supercomputers.

Although we have been unable to use entropy to separate complexity classes due to the fact that entropy only captures limiting behavior of a dynamical system, the dynamical system implementation on the continuum reveals a relationship between two concepts so far apparently unrelated: nondeterminism and entropy. The entropy is an invariant of a dynamical implementation of a deterministic automaton. (For nondeterministic automata, this seems to be an open question.) Whether this relation can be used to separate deterministic from nondeterministic complexity classes is a question worthy of further research.

A second question concerns the impact of this notion of implementation of concurrent computation. It is conjectured in [2] that piecewise linear systems on euclidean spaces *cannot* implement synchronous systems such as cellular automata in any dimension. If true, the conjecture raises the question whether it is possible at all to implement synchronous and/or asynchronous concurrent systems by piecewise linear dynamical systems in euclidean spaces.

Lastly, along similar arguments, euclidean spaces can be distinguished by the possibility of implementing sequential automata of a given complexity. (One-dimensional euclidean spaces do not allow implementation of Turing machines but that is possible in dimension 2, where cellular automata may not be implementable.) Can they be separated analogously by the concurrent devices that they carry within?

References

1. R. Bartlett, M. Garzon: Computational Complexity of Maps of the Interval. Technical Report, The University of Memphis, 1996.
2. M. Cosnard, M. Garzon, P. Koiran: Computability properties of low-dimensional dynamical systems. Ext. Abs. in *Lecture Notes in Computer Science* **665** (1993). Final version in *Theoret. Comp. Sci.* **132** (1994) 113–128.
3. R.L. Devaney: An Introduction to Chaotic Dynamical Systems. Addison-Wesley, Reading MA, (1986).
4. J. Hopcroft and J. Ullman: *An Introduction ot the Theory of Computation: Automata, Languages and Machines.* Addison-Wesley (1979).
5. J. Milnor and W. Thurston: On Iterated Maps of the Interval. In: J. C. Alexander (Ed.): *Dynamical Systems*: Lecture Notes in Mathematics **1342**, A. Dold and B. Eckmann (Eds.) (1987).

6. M. Garzon: *Models of Massive Parallelism: Analysis of Cellular Automata and Neural Networks*. Springer-Verlag, 1995.
7. C. Moore: Generalized One-sided Shifts and maps of the interval. *Nonlinearity* 4 (1991) 727–745.
8. C. Moore: Dynamical Recognizers: Real-time Language Recognition by Iterated Maps. Technical Report, Santa Fe Institute, 1996.
9. J.L. Peterson: *Petri Net Theory and the Modeling of Systems*, Prentice-Hall, 1981.
10. H.T. Siegelmann, E.D. Sontag: *On the computational power of neural nets.* SYCON Report 91–11, Rutgers University, 1991.
11. S.M. Shatz: *Development of Distributed Software: Concepts and Tools*, Macmillan, 1993.

Implementing WS1S via Finite Automata

James Glenn[1] and William Gasarch[2]

[1] Department of Computer Science, University of Maryland, College Park MD 20742,
USA
[2] Department of Computer Science and Institute for Advanced Computer Studies,
University of Maryland, College Park MD 20742, USA

Abstract. It has long been known that $WS1S$ is decidable through the
use of finite automata. However, since the worst-case running time has
been proven to grow extremely quickly, few have explored the implementation of the algorithm. In this paper we describe some of the points
of interest that have come up while coding and running the algorithm.
These points include the data structures used as well as the special features of the automata, which we can exploit to perform minimization
very quickly in certain cases. We also present some data that enable us
to gain insight into how the algorithm performs in the average case, both
on random inputs and on inputs that come from the use of Presburger
Arithmetic (which can be converted to $WS1S$) in compiler optimization.

1 Introduction

In the study of theoretical computer science, we are often concerned with whether
or not a problem can be solved, and if so, what is the worst case running time. In
situations where the worst-case has been shown to be very bad, it is often the case
that the problem is not pursued any further. Any issues, such as performance
or data structure issues, that may be encountered while trying to implement a
solution are left unknown and undealt with. We have implemented an algorithm
for one such problem: the decision procedure for $WS1S$, the weak second-order
theory of the natural numbers with $<$ and successor, where quantified variables
range over $\mathbf{N}$ and finite subsets of $\mathbf{N}$ (as opposed to the strong second-order
theory, in which quantified variables can range over infinite subsets as well). In
the process, we have come across many problems that are worth investigation.
One problem with which we must deal is how to keep the size of our automata
from growing too fast, and how to do so efficiently. This led us to employ two
new strategies for minimization that rely on the structure of the automata we
build. One of these methods is a new algorithm that minimizes certain automata
in linear time.

We also hope to show that, in practice, the problem is not as hard as the
worst case would lead us to believe. One difficulty with this hope is the notion
of "in practice". As our "practical" input, we shall use both random data and
data from real world applications. The latter is possible because we now have
several inputs from the use of Presburger Arithmetic by compilers that solve
dependency problems to transform parallel loops. It can be shown that any

Presburger formula can be converted to a formula in L_{S1S}. We also propose to implement algorithms for other problems, such as $WS2S$, with the hope that we will uncover interesting topics there as well.

2 $WS1S$

2.1 Definition

In the language L_{S1S}, the relation symbols are $<$ and $\in$, the only function symbol is the one-place function s (meaning successor (and with $s^n(x)$ written as $x+n$)), and the constants are $0, 1, \ldots$ and sets of those numbers. We need not concern ourselves with the relation $=$ because (under the standard interpretation) $x = n$ can be written as $\neg(x < n \vee n < x)$, while $X = S$ can be written as $\forall z(z \in X \iff z \in S)$. This gives rise to atomic formulas of the following forms:

(1) $c_1 < c_2$
(2) $c_1 < x_1 + c_2$
(3) $x_1 + c_1 < c_2$
(4) $x_1 + c_1 < x_2 + c_2$
(5) $c_1 \in S$
(6) $x_1 + c_1 \in X_1$
(7) $x_1 + c_1 \in S$
(8) $x_1 + c_1 \in X_1$

where c_1 and c_2 are natural numbers, x_1 and x_2 are scalar variables, X_1 is a finite set variable, and S is a finite set constant (as they will be throughout this paper).

Note that formulas of form (1) can be evaluated with no difficulty, and that those of form (5) can be rewritten in terms of $=$ (and hence $<$):

$$x_1 \in \{c_1, \ldots, c_n\} \iff (x_1 = c_1 \vee \cdots \vee x_1 = c_n),$$

so we will not concern ourselves with those forms; we assume, without loss of generality, that formulas are built from atomic formulas of forms (2), (3), (5), (6), (7), and (8). Our procedure then works on sentences such as the following:

$$\forall X \exists Y \forall z(\neg(z \in X) \vee (z + 1 \in Y))$$
$$\exists X \forall y(y \in X).$$

The set $WS1S$ is the set of sentences of L_{S1S} which are true under the standard interpretation. Hence, of the above sentences, the first is in $WS1S$ but the second is not (remember that quantifiers in a weak second-order language are over finite sets).

2.2 Decidability of WS_1S

Büchi showed in 1960 that $WS1S$ is decidable through the use of finite automata [1]. However, the cost of the decision procedure can be very high. In fact, it has been proven that for all sufficiently large n, there is a sentence of length n that requires time proportional to $t_{\epsilon \log_2 n}(n)$, where t_k is the tower function with height k [3].

We present a full proof of decidability again in order to indicate where performance issues arise. Consider a formula f of one free variable, either set or scalar. We will demonstrate how to construct a deterministic finite automaton that corresponds to f. Naturally there will be more than one such automaton, and we will call the set of such automata $DFA(f)$ with the idea being that if $M \in DFA(f)$ then M accepts only those strings of zeros and ones that are encodings of solutions to f. For formulas with more than one free variable the idea will be similar, except that M will operate on a multi-track tape with each variable written on a different track of the tape. The encoding of a set X has a one in place n if $n \in X$ and a zero otherwise. The encoding of a natural number n will be the empty string if $n = 0$; otherwise it will consist of $n - 1$ zeros followed by a one. Note that this encoding does not assign meaning to every tape, in particular to those with digits after a one. We do, however, wish do give these tapes meaning, and we will do so after the following definitions.

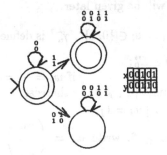

Fig. 1. An automaton in $DFA(x = y)$ and an input it accepts.

Definition 1. The empty string is denoted e.

Definition 2. The symbol on a zero track tape will be denoted $\perp$.

Definition 3. $[N]^{<\omega}$ is the set of finite subsets of N.

Definition 4. $\Sigma_0 = \{\perp\}$, $\Sigma_1 = \{0,1\}$, $\Sigma_2 = \{0,1\} \times \{0,1\}$ and in general $\Sigma_n = \Sigma_{n-1} \times \{0,1\}$. These are the alphabets of our automata. We will write the symbols in these alphabets as stacks or column or row vectors as is convenient. Also, define $Zero_n$ to be the element of Σ_n all of whose components are zero.

Definition 5. Let γ, our encoding function, which maps $\bigcup_{k=0}^{\infty}(\mathbf{N} \cup [\mathbf{N}]^{<\omega})^k$ (tuples whose components are natural numbers or finite sets of natural numbers) to $\bigcup_{k=0}^{\infty} \Sigma_k^*$ be defined by

(1) $\gamma() = \perp$;

(2) $\gamma(\alpha) = \begin{cases} e & \text{if } c_1 = 0 \\ 0^{\alpha-1}1 & \text{otherwise} \end{cases}$ if $\alpha \in \mathbf{N}$;

(3) $\gamma(\alpha) = w_0 \cdots w_k$ where $w_i = 1$ if $i \in \alpha$ and $w_i = 0$ otherwise and $k = \max_{i \in \alpha} i$ whenever $\alpha \in [\mathbf{N}]^{<\omega}$;

(4) for $(\alpha_1, \ldots, \alpha_n)$, $n \geq 2$ write $w = w_1 \ldots w_k$ for $\gamma(\alpha_1, \ldots, \alpha_{n-1})$ and $u_1 \ldots u_l$ for $\gamma(\alpha_n)$. Extend w or u with $Zero_{n-1}$ or $Zero_1$ respectively so they have the same length. Then for $1 \leq i \leq \max\{k, l\}$ let $v_i = \begin{pmatrix} w_i \\ u_i \end{pmatrix}$ (that is, v_i is the element of Σ_n that is equal to w_i in its top $n - 1$ rows and has u_i in its bottom row). Then let $\gamma(\alpha_1, \ldots, \alpha_n) = v_1 \ldots v_{\max\{k,l\}}$.

Definition 6. $M \in DFA(f)$ if and only if

(1) M accepts w implies $f(\gamma^{-1}(w))$ is true; and
(2) $f(\bar{x})$ is true implies M accepts $\gamma(\bar{x})$.

In the above definition of $DFA(f)$ we must be careful since the encoding function is not onto, and hence its inverse will not be defined for all strings w. We therefore desire an extended inverse function defined over all of $\bigcup_{k=0}^{\infty} \Sigma_k^*$; more motivation for this will be given later.

Definition 7. Let $u = u_1 \ldots u_l \in \{0,1\}^*$. γ_u^{-1} is defined as follows:

(1) $\gamma_e^{-1}(w) = ()$ for all $w \in \Sigma_0^*$;

(2) $\gamma_0^{-1}(w_1 \ldots w_k) = \begin{cases} 0 & \text{if } w_1 = \cdots = w_k = 0 \\ \min\{i \mid w_i \neq 0\} & \text{otherwise} \end{cases}$;

(3) $\gamma_1^{-1}(w_1 \ldots w_k) = \{i - 1 \mid w_i = 1\}$;

(4) If $w_1 \ldots w_k \in \Sigma_n^*$, $n \geq 2$, write $w_i = \begin{pmatrix} w_{i1} \\ \vdots \\ w_{in} \end{pmatrix}$ for $1 \leq i \leq k$ and let

$$\gamma_u^{-1}(w_1 \ldots w_k) = (\gamma_{u_1}^{-1}(w_{11} \ldots w_{k1}), \ldots, \gamma_{u_n}^{-1}(w_{1n} \ldots w_{kn})).$$

Note that we had to define γ^{-1} as a family of functions since γ_0^{-1} and γ_1^{-1} have the same domain but behave differently. This distinction is not important and will be omitted; we will write γ^{-1} whenever we mean one of the γ_u^{-1}'s. Finally, we offer some examples and a lemma.

$$\gamma(\emptyset) = \gamma_{1,0}(0) = e$$
$$\gamma(3) = 001$$
$$\gamma(\{4,6\}) = 0000101$$
$$\gamma(3, 1, \{1, 2\}) = \begin{matrix} 0 & 0 & 1 \\ 1 & 0 & 0 \\ 0 & 1 & 1 \end{matrix}$$
$$\gamma^{-1}(\perp\perp\perp) = ()$$

$$\gamma_{00}^{-1}\begin{pmatrix} 0\,0\,0\,0 \\ 0\,0\,0\,0 \end{pmatrix} = (0,0)$$

$$\gamma_{10}^{-1}\begin{pmatrix} 0\,0\,0\,0 \\ 0\,0\,0\,0 \end{pmatrix} = (\emptyset,0)$$

$$\gamma_{110}^{-1}\begin{pmatrix} 0\,0\,1\,0 \\ 1\,0\,1\,1 \\ 1\,0\,1\,0 \end{pmatrix} = (\{2\},\{0,2,3\},1)$$

Lemma 8. *If $w \in \Sigma_n^*$ and $u \in Zero_n^*$ then $\gamma^{-1}(wu) = \gamma^{-1}(w)$.*

Automata Corresponding to Atomic Formulas. We are now ready to describe how to construct an automaton corresponding to any formula f. We will work on a formula f from inside out. That is, we will construct automata for the atomic formulas from which f is built and combine them according to the logical connectives and quantifiers between them.

Let $n, m \in \mathbf{N}$ and the formula $f(x)$ be $n < x + m$. We can, without loss of generality, assume that $m = 0$ and $n > 0$, since if $n < m$ the formula is true for all x, otherwise f is equivalent to $(n - m) < x$. In this case, if $n = m$ then $f(x)$ is true for all $x > 0$ and we build an automaton that rejects only 0^*. If $n > m$ we build a (nondeterministic) automaton (Figure 2) that accepts w if and only if $\gamma^{-1}(w) = x$ and $n < x$.

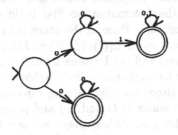

Fig. 2. Automaton for $n < x$

This automaton accepts $0^* \cup 0^n 0^* 1\{0,1\}^*$. If $w \in 0^*$ then $\gamma^{-1}(w) = 0$ and since $0 < n$ if $n > 0$ then w is as required. If $w \in 0^n 0^* 1\{0,1\}^*$ then $w = 0^k 1u$ for some $k \geq n$ and $u \in \{0,1\}^*$. Now by the definition of the extended inverse function and our lemma, $\gamma^{-1}(w) = \gamma^{-1}(0^k 1) = k + 1$. So $k > n$ and w is again as required. It is easy to check that if $f(x)$ is true then $\gamma(x)$ is accepted by the automaton. Note that while the automaton displayed is nondeterministic, we can convert it to a deterministic automaton.

Now let $m \in \mathbf{N}$ and consider the formula $y + m \in X$ which will be denoted $f(X, y)$. We build the finite automaton (Figure 3) that accepts

$$\begin{smallmatrix}1\\0\end{smallmatrix}\begin{pmatrix}\begin{smallmatrix}0\\0\end{smallmatrix} \cup \begin{smallmatrix}1\\0\end{smallmatrix}\end{pmatrix}^* \cup \begin{pmatrix}\begin{smallmatrix}0\\0\end{smallmatrix} \cup \begin{smallmatrix}1\\0\end{smallmatrix}\end{pmatrix}^* \begin{pmatrix}\begin{smallmatrix}0\\1\end{smallmatrix} \cup \begin{smallmatrix}1\\1\end{smallmatrix}\end{pmatrix}\begin{pmatrix}\begin{smallmatrix}0\\0\end{smallmatrix} \cup \begin{smallmatrix}0\\1\end{smallmatrix} \cup \begin{smallmatrix}1\\0\end{smallmatrix} \cup \begin{smallmatrix}1\\1\end{smallmatrix}\end{pmatrix}^m \begin{pmatrix}\begin{smallmatrix}1\\0\end{smallmatrix} \cup \begin{smallmatrix}1\\1\end{smallmatrix}\end{pmatrix}\begin{pmatrix}\begin{smallmatrix}0\\0\end{smallmatrix} \cup \begin{smallmatrix}0\\1\end{smallmatrix} \cup \begin{smallmatrix}1\\0\end{smallmatrix} \cup \begin{smallmatrix}1\\1\end{smallmatrix}\end{pmatrix}^*.$$

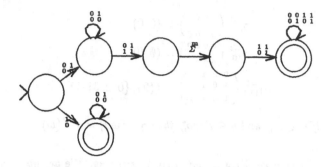

Fig. 3. Automaton for $y + m \in X$

This automaton then accepts w if and only if $\gamma^{-1}(w) = (X, y)$ and $f(X, y)$ is true.

We can also display automata for atomic formulas of forms (3), (5), (6), (7), and (8); this is left to the full paper.

Automata for Non-atomic Formulas. Constructing an automaton for a formula of the form $f(\bar{x}) = g(\bar{x}) \wedge h(\bar{x})$ or $f(\bar{x}) = g(\bar{x}) \vee h(\bar{x})$ is easy, but does require the use of a data structure that allows us to insert and retrieve ordered pairs quickly. To build the automaton, we first build automata $G \in DFA(g)$ and $H \in DFA(h)$ and combine them in the standard cross-product way to get an automaton F accepting either $L(G) \cap L(H)$ or $L(G) \cup L(H)$. The proof that $F \in DFA(f)$ is simple and will not be done here. Of course, some of the states produced for F may not be reachable; it is desirable when implementing this algorithm to ignore them altogether. By building F forward from (s_G, s_H) (where s_G and s_H are the start states of G and H) and proceeding only to reachable states, we can achieve this goal. This strategy requires that we are able to keep track of what states we have constructed so far; we use an xy-tree for this purpose. Elaboration of this, along with that of other implementation details, will be left for the full paper.

For a formula of the form $f(\bar{x}) = \neg g(\bar{x})$ we can construct $G \in DFA(g)$ and make its final states non-final and vice-versa to get F. As above, the proof that the new automaton is in $DFA(f)$ is simple and not expounded here.

Finally, we have formulas involving quantifiers – those of the form

$$h(\alpha_{i_1}, \ldots, \alpha_{i_{n-1}}) = \exists \alpha_{i_n} g(\alpha_{i_1}, \ldots, \alpha_{i_n})$$

or

$$h(\alpha_{i_1}, \ldots, \alpha_{i_{n-1}}) = \forall \alpha_{i_n} g(\alpha_{i_1}, \ldots, \alpha_{i_n}),$$

where α_{i_n} is free in g. Since we can write $\forall \alpha_{i_n} g(\alpha_{i_1}, \ldots, \alpha_{i_n})$ as $\neg \exists \alpha_{i_n} \neg g(\alpha_{i_1}, \ldots, \alpha_{i_n})$, we can assume, without loss of generality, that all quantifiers are existential. Let $G \in DFA(g)$ and $G = (K, \Sigma_n, \delta, s, F)$. Then let $H = (K, \Sigma_{n-1}, \Delta', s, F')$ be a nondeterministic finite automaton (which, of course, we can make deterministic) and let

$$\Delta' = \{(q, (\sigma_1, \ldots, \sigma_{n-1}), q') \mid \exists b \in \{0,1\} \ni \delta(q, (\sigma_1, \ldots, \sigma_{n-1}, b)) = q'\}$$

$$F' = \{q \in Q \mid \exists w \in Zero_{n-1}^*, f \in H \ni (q, w) \vdash_H^* (f, e)\}.$$

Theorem 9. $H \in DFA(h)$.

Proof. Fix $\alpha_{i_1}, \ldots, \alpha_{i_{n-1}}$ and suppose that $h(\alpha_{i_1}, \ldots, \alpha_{i_{n-1}})$ is true. That means that there exists an α_{i_n} such that $g(\alpha_{i_1}, \ldots, \alpha_{i_n})$ is true. Let

$$\gamma(\alpha_{i_1}, \ldots, \alpha_{i_n}) = w_1 \ldots w_m = \begin{pmatrix} b_{11} \\ \vdots \\ b_{1n} \end{pmatrix} \cdots \begin{pmatrix} b_{m1} \\ \vdots \\ b_{mn} \end{pmatrix}$$

and

$$u = u_1 \ldots u_m = \begin{pmatrix} b_{11} \\ \vdots \\ b_{1(n-1)} \end{pmatrix} \cdots \begin{pmatrix} b_{m1} \\ \vdots \\ b_{m(n-1)} \end{pmatrix},$$

that is u is w with the bottom track removed. We know G must accept w. We need to show that H accepts $\gamma(\alpha_{i_1}, \ldots, \alpha_{i_{n-1}})$. Since G accepts w, there must exist $q_0, \ldots, q_m \in K$ with $q_0 = s$ and $q_m \in F$ such that

$$(q_0, w_1 \ldots w_m) \vdash_G (q_1, w_2 \ldots w_m) \vdash_G \cdots \vdash_G (q_{m-1}, w_m) \vdash_G (q_m, e).$$

So for $0 \le i < m$, $\delta(q_i, (b_{i+1,1}, \ldots, b_{i+1,n})) = q_{i+1}$. This means that, for $0 \le i < m$ there exists a $b \in \{0,1\}$ (namely $b_{i+1,n}$) such that $\delta(q_i, \begin{pmatrix} b_{i1} \\ \vdots \\ b_{i(n-1)} \\ b_i \end{pmatrix}) = q_{i+1}$.

This means that for each i, $0 \le i < m$

$$(q_i, \begin{pmatrix} b_{i1} \\ \vdots \\ b_{i(n-1)} \end{pmatrix}, q_{i+1}) \in \Delta'.$$

Therefore,

$$(q_0, u_1 \ldots u_m) \vdash_H^* (q_1, u_2 \ldots u_m) \vdash_H^* \cdots \vdash_H^* (q_{m-1}, u_m) \vdash_H^* (q_m, e)$$

so H accepts u. However $\gamma(\alpha_{i_1}, \ldots, \alpha_{i_{n-1}})$ will equal some prefix of u, $u_1 \ldots u_l$. Then $u_{l+1} = \cdots = u_m = Zero_{n-1}$. So

$$(q_0, u_1 \ldots u_l) \vdash_H^* (q_l, e)$$

and

$$(q_l, Zero_{n-1}^{m-l}) \vdash_H^* (q_m, e).$$

So by the construction of F' we have $q_l \in F'$ so we say that H accepts $u_1 \ldots u_l$, which is what needed to be shown.

Now suppose that H accepts

$$u = u_1 \ldots u_m = \begin{pmatrix} b_{11} \\ \vdots \\ b_{1(n-1)} \end{pmatrix} \cdots \begin{pmatrix} b_{m1} \\ \vdots \\ b_{m(n-1)} \end{pmatrix}.$$

We need to show that $h(\gamma^{-1}(u))$ is true. We know that

$$(q_0, u_1 \ldots u_m) \vdash_H^* (q_1, u_2 \ldots u_m) \vdash_H^* \cdots \vdash_H^* (q_{m-1}, u_m) \vdash_H^* (q_m, e)$$

where $q_0 = s_F$ and $q_m \in F'$. Furthermore, by the construction of F' we know that for some $u' \in u \cdot Zero_{n-1}^*$

$$(q_0, u_1' \ldots u_{m'}') \vdash_H^* (q_1, u_2' \ldots u_{m'}') \vdash_H^* \cdots \vdash_H^* (q_{m'-1}, u_{m'}') \vdash_H^* (q_{m'}, e)$$

where $q_{m'} \in F$. So for each $0 \leq i < m'$, $(q_i, u_{i+1}', q_{i+1}) \in \Delta'$. If we write

$$u_i' = \begin{pmatrix} b_{i1} \\ \vdots \\ b_{i(n-1)} \end{pmatrix}$$

for $i > m$ and appeal to the definition of Δ' we see that for each i, $0 \leq i < m'$, there exists a b_i such that $\delta(q_i, (b_{i,1}, \ldots, b_{i,n-1}, b_i)) = q_{i+1}$. We write

$$w_i' = \begin{pmatrix} b_{i1} \\ b_{i(n-1)} \\ b_i \end{pmatrix}$$

and then see that

$$(q_0, w_1' \ldots w_{m'}') \vdash_G^* (q_{m'}, e).$$

So G accepts $w' = w_1' \ldots w_{m'}'$ and $g(\gamma^{-1}(w'))$ must be true. But if $\gamma^{-1}(w') = (\alpha_{i_1}, \ldots, \alpha_{i_n})$, then for $(\alpha_{i_1}, \ldots, \alpha_{i_{n-1}})$ there exists an α_{i_n} such that $g(\alpha_{i_1}, \ldots, \alpha_{i_n})$ is true, in other words, $h(\alpha_{i_1}, \ldots, \alpha_{i_{n-1}})$ is true. Now note that $\gamma^{-1}(u') = (\alpha_{i_1}, \ldots, \alpha_{i_{n-1}})$ since u' is just w' with the last track removed. So $h(\gamma^{-1}(u'))$ is true. Finally, since $u' \in u \cdot Zero_{n-1}^*$ we have by a previous lemma $\gamma^{-1}(u') = \gamma^{-1}(u)$. $h(\gamma^{-1}(u))$ is then true as well. $\square$

Three things are worth noting here. First, if the domain of γ^{-1} had not been all of Σ_n^* we would have run into trouble – when we do the nondeterminism modification the new automaton accepts those strings w for which we can add a new track to get a string u accepted by G. What if the new track had no inverse? We would be saying that $f(\alpha_{i_1}, \ldots, \alpha_{i_{n-1}})$ is true because there is an α_{i_n}, namely *garbage*, such that $g(\alpha_{i_1}, \ldots, \alpha_{i_n})$ is true, which is clearly not what we want. We could have addressed this problem by generating a machine that accepts strings w such that you can add a new track b to get a string accepted by G and b has an inverse. We chose to solve the problem by making $domain(\gamma^{-1}) = \Sigma_n^*$ simply because it seemed easier to implement.

Second is the necessity of modifying the set of final states, which is best illustrated by an example (Figure 4). The last automaton rejects e when it should accept all input. Since final state AB is reachable from state A on input 0, we should have made A a final state too. It is easy to check that doing so does indeed result in an automaton that accepts all input.

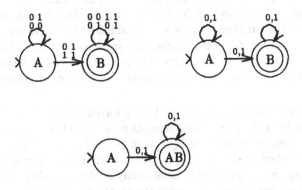

Fig. 4. Result of erasing a track without altering final states

Third are some comments on implementation. We could create a nondeterministic finite automaton from a deterministic one, and then send the result to a procedure to make it deterministic again. This would require the use of a new NFA data structure which would duplicate much of the DFA code. We can, however, do away altogether with the need for this separate type by combining the two routines into one: we send in a deterministic automaton and get one back. This proves to be simple to do. Again, discussion is postponed until the final paper.

Now for any sentence $f(\bar{x})$ we can construct the corresponding machine F. It is easy to check whether F accepts Σ_0^*. If it does, then $f \in WS1S$, otherwise it is not. However, it may not be easy to build the automaton corresponding to f. For each quantifier in the formula we construct a nondeterministic automaton. Most of the algorithms mentioned so far could be modified to work on nondeterministic automata, except for the algorithm for $\neg f$. There, the input needs to be deterministic for the output to be correct. The algorithm for converting an NFA to a DFA works in worst-case $O(2^{|K|})$ time, and hence for each quantifier the size of the machine (and the time needed to create it) may increase exponentially (this, however, should be unlikely).

2.3 Minimization

Common sense tells us that to keep the running time of our algorithm down, we should periodically minimize our automata. If we keep around machines with twice as many states as necessary, not only will we run out of memory faster, but the and/or algorithm will take four times as long to run. The nondeterminism routine will be even worse, for a machine's bloat will increase exponentially. Empirical results tell us that minimization is not only desirable, but practically mandatory – it is easy to find inputs that will run out of memory on common workstations. These studies also tell us that we should minimize our automata at almost every opportunity. Because minimizing automata is so important (in addition to being interesting in its own right), we have looked at three ways of approaching the problem. One is well established in the literature, and two are new strategies we developed during the implementation of the algorithm. Of these, one is easy to do but gives only partial results, and the other is an algorithm that exploits an interesting property of certain automata.

Trap State Reduction. Suppose we have a formula of the form $f(\bar{x}) = g(\bar{x}) \wedge h(\bar{x})$ and let q be a state of H that accepts nothing. Then our new automaton F will have a possibly large set of equivalent states $\{(q, q') \mid q' \in G\}$ that accept nothing. If we detect that we have generated a state (q_G, q_H) such that q_G or q_H accepts nothing, we do not have to explore any of the possible successors of that state, and if we generate (q'_G, q'_H) with the same property, we can immediately mark it as equivalent to (q_G, q_H). This strategy will work in the nondeterminism algorithm as well. Empirical data show that though this is helpful, it alone will not solve the problem of large machines.

Hopcroft's Algorithm. We also employ a general-purpose automaton minimization algorithm. Hopcroft presented in 1976 an algorithm that works in $O(|\Sigma| n\log n)$ time in the worst case [2]. Empirical results show that the constant factors are such that it easily beats the standard $O(|\Sigma| n^2)$ algorithm even on small machines.

Layered Automata. Finally, we can exploit the special structure of some of our automata to use a new algorithm that runs in $O(|\Sigma| n)$ time. The algorithm is presented below. The special requirement is that the automaton has no non-trivial cycles (formally, if $(q_0, w_0) \vdash (q_1, w_1) \vdash \cdots \vdash (q_n, w_n) \vdash (q_0, w_{n+1})$ then $q_0 = q_1 = \cdots = q_n$). All of the automata generated for atomic formulas are of this structure. It is preserved through the operations of negation, and, and or. Unfortunately, existential quantifiers can destroy the structure (Figure 5).

Preliminary tests show that on average, our algorithm works almost twice as fast as Hopcroft's. It is therefore worthwhile to use it whenever possible; that is on subformulas with no quantifiers. If the formula has all its quantifiers in front, this is nearly every subformula.

The algorithm is as follows:

(1) For each state q, initialize $votes_q$ to $|\{\sigma \mid \delta(q, \sigma) = q\}|$.
(2) Divide those states with $votes_q = |\Sigma|$ into final and non-final; mark each member of these groups as equivalent to the others, mark them all as level 0 and let $currentlevel = 1$.

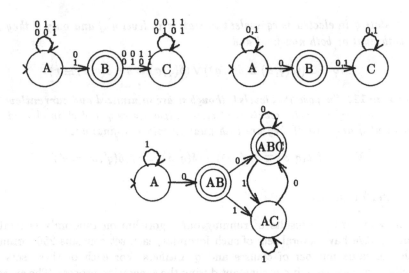

Fig. 5. Layers are not preserved by nondeterminism

(3) For each state q in level 0 and $\sigma \in \Sigma$, if $\delta(q, \sigma) = q'$ then increment $votes_{q'}$.

(4) Collect all states with $votes_q = | \Sigma |$ and call them *elected*. If *elected* is empty, then halt.

(5) If any state q in *elected* is equivalent to a state in level $currentlevel - 1$ mark them as equivalent, move it from *elected* to level $currentlevel - 1$, and for each $\sigma \in \Sigma$ if $\delta(q, \sigma) = q'$ then increment $votes_{q'}$, adding q' to *elected* if $votes_{q'} = | \Sigma |$.

(6) If any states were added to *elected* by the previous step, then repeat it.

(7) Mark those states remaining in *elected* as in level $currentlevel$ and check for equivalence among them.

(8) Elect new states into *elected* as in steps 3 and 5, increment $currentlevel$, and return to step 4.

The following lemmas provide the basis of a correctness proof of the above algorithm.

Theorem 10. *Suppose that levels 0 through n are minimized and $currentlevel = n + 1$. Then the following hold before each execution of step 5:*

(a) Any state in elected has at least one transition into level n.

(b) No state in elected has a transition to a different state in elected

(c) No state in elected is equivalent to any state in levels 0 to $n - 1$.

(d) If a state q in elected has $delta(q, \sigma) = r$ for some $\sigma \in \Sigma$ and r in level n, and $s \neq r$ is also in level n, then q is not equivalent to s.

(e) A state q in elected is equivalent to state r in level n if and only if they are both final or both non-final and

$$\forall \sigma \in \Sigma((\delta(q,\sigma) \equiv \delta(r,\sigma)) \vee (\delta(q,\sigma) = q \wedge \delta(r,\sigma) \equiv r))$$

Theorem 11. *Suppose that levels 0 though n are minimized and currentlevel = $n+1$. Then before each execution of step 7, two states q and q' in elected are equivalent if and only if they are both final or both non-final and*

$$\forall \sigma \in \Sigma((\delta(q,\sigma) \equiv \delta(q',\sigma)) \vee (\delta(q,\sigma) = q \wedge \delta(q',\sigma) = q'))$$

2.4 Performance

We now present the results of running our algorithm on randomly generated formulas. We have several sets of such formulas, each set contains 250 formulas with the same number of clauses and quantifiers. For each of these sets we specified the same maximum constant during the generation process. The average times (in milliseconds) for each set are given below.

quantifiers	2		3		4	
clauses	8	16	8	16	8	16
$max = 2$	33	57	45	86	91	159
$max = 3$	40	70	59	111	133	226
$max = 4$	46	85	79	143	210	335
$max = 5$	56	100	99	183	289	517
$max = 6$	67	122	122	236	418	728
$max = 7$	78	140	175	297	632	1037
$max = 8$	99	177	294	547	1213	1640

¿From this data we can see that adding a quantifier increases running time by a factor of about two. Also, the effect of increasing the size of the constants is much more profound on formulas with more quantifiers. We should also note that for a randomly selected sample of formulas from this test we ran the decision procedure with Hopcroft's algorithm substituted for the layered automaton minimization procedure. The performance was degraded by an average of ten percent.

3 Presburger Arithmetic

Presburger Arithmetic, the first-order theory of the integers with $+$ and $<$, is decidable. There is an algorithm to determine the truth of a Presburger formula in $2^{2^{2^{pn \log n}}}$ time [4] which works by eliminating quantifiers, converting infinite searches to finite searches. At the end, all the algorithm has to do is check the finite (but very large) number of cases. Another approach to the problem is to convert a Presburger Formula into $WS1S$ and run the algorithm described in the last section on that formula. We do not expect this method to provide a faster algorithm for deciding Presburger Arithmetic since the bound for $WS1S$ is even worse that that for Presburger Arithmetic, but the formulas that result provide us with a source of real-world test inputs (Presburger Formulas are used by compilers that optimize loops for parallel execution). We hope to show that we can verify reasonable Presburger formulas in a reasonable amount of time.

In the conversion from Presburger Arithmetic to $WS1S$, the integer variables and terms become set variables with a set variable representing an integer if its characteristic function, written out as a string, is essentially the binary representation of the integer. This conversion is well-known; we will omit the details.

We have converted a random set of Presburger formulas to L_{S1S} and run the $WS1S$ decision procedure on them. The random formulas we generated were all of a simple form, so the raw results below will most likely not hold for more general formulas. However, the *trends* shown below may still appear in the general case.

quantifiers	maximum constant		
	4	8	16
2	2.21	3.21	10.85
3	10.45	28.08	
4	68.03		

This table (for which we are still gathering data) gives average times (in seconds) for deciding formulas with a given number of variables and a given bound on the constants that appear in the formula. We can see that changing either parameter has a such profound effect on the running time that filling out this table even a few rows or columns past its current boundary will be quite time consuming.

4 Future Studies

As stated before, we would like to find some conditions under which we can get a more reasonable upper bound on the running time for the $WS1S$ decision procedure. Perhaps we can exploit the layered structure of the automata for this purpose.

Many of the choices we made for our implementation need to be examined, among them our tape encoding, minimization algorithms, and the way in which we deal with quantifiers. In the case of the tape encoding, we wish to find if a different encoding could lead to improved running time; the special case for zero in our current encoding is particularly cumbersome. We have already discussed briefly the results of some tests of minimization algorithms, however our study of such algorithms has been by no means comprehensive. In particular, Brzozowski's algorithm is acclaimed in the literature, yet we have not studied how it performs on our automata. Indeed, the automata corresponding to $WS1S$ formulas may provide a useful source of test input for future studies of minimization algorithm performance. As for quantifiers, it should be pointed out that if the same quantifier occurs k times consecutively, we could erase k tracks at once, thus requiring only one NFA to DFA conversion, but with the trade-off that for each state we need to see where it goes on 2^k inputs, rather than just two. If this is an effective way to handle quantifiers, it would be especially important for Presburger formulas, which as noted above require a lot of quantifiers to be converted to L_{WS1S}. Also, in these areas and undoubtedly in many others we shall explore the use of different data structures to the reduce run-time and storage requirements of the algorithm, which are currently quite high.

Finally, we propose to implement the decision procedure for $WS2S$, which uses tree automata instead of DFA's. In doing so we will address some of the same issues as for $WS1S$. For example, it is currently unknown whether there is an analog of Hopcroft's minimization algorithm for tree automata. We can also hope that most of the tree automata we build have a structure similar to the monotone structure of many of our DFA's.

References

1. J. R. Büchi. Weak second order arithmetic and finite automata. *Zeitscrift fur mathematische Logic und Grundlagen der Mathematik*, 6:66–92, 1960.
2. J. Hopcroft. An n log n algorithm for minimizing states in a finite automaton. In Z. Kohavi and A. Paz, editors, *Theory of Machines and Computation*, pages 189–196. Academic Press, 1976.
3. A. R. Meyer. Weak monadic second order theory of successor is not elementary-recursive. In *Logic Colloquium*, number 453 in Lecture Notes in Mathematics, pages 132–154. Springer-Verlag, 1974.
4. D. C. Oppen. Elementary bounds for Presburger arithmetic. In *5th ACM Symposium on Theory of Computing*, pages 34–37, 1973.

Acknowledgements

William Gasarch is supported in part by NSF grants CCR-8803641 and CCR-9020079.

Instruction Computation in Subset Construction*

J. Howard Johnson[1] and Derick Wood[2]

[1] Institute for Information Technology, National Research Council, Nepean, Ontario,
Canada. E-mail: johnson@iit.nrc.ca.
[2] Department of Computer Science, Hong Kong University of Science & Technology,
Clear Water Bay, Kowloon, Hong Kong. E-mail: dwood@cs.ust.hk.

Abstract. Subset construction is *the* method of converting a nondeterministic finite-state machine into a deterministic one. The process of
determinization is important in any implementation of finite-state machines. The reasons are that nondeterministic machines are often easier
to describe than their deterministic equivalents and the conversion of
regular expressions to finite-state machines usually produces nondeterministic machines.

We discuss one aspect of subset construction; namely, the computation
of the instructions of the equivalent deterministic machine. Although the
discussion is to a large extent independent of any specific assumptions,
we draw some conclusions within the context of INR and Grail, both
of which are packages for the manipulation of finite-state machines. Related work is described by Aho [1] and by Crochemore and Rytter [4] in
the context of pattern matching, and by Perrin [10] in the context the
conversion of regular expressions into finite-state machines.

The aim of the discussion is to present the problem and suggest some
possible solutions.

1 Introduction

Subset construction is *the* method of converting a nondeterministic finite-state
machine into a deterministic one. The basic idea underlying the conversion is
the use of sets of states (state sets) of the nondeterministic finite-state machine
as states in the corresponding equivalent deterministic finite-state machine. It
is simple, yet nontrivial, to implement well. One immediate practical problem
is that, for a given nondeterministic finite-state machine that has n states, the
resulting deterministic finite-state machine may have $2^n - 1$ states. Moreover,
there are nondeterministic finite-state machines that yield $\Omega(2^n)$ states. Thus,
the exponential growth factor cannot be avoided in general. There are two possible avenues, at least, of investigation that this bleak news suggests: to improve
the efficiency of determinization under the assumption that exponential blow up

* This work was supported under grants from the Natural Sciences and Engineering
Research Council of Canada, from the Information Technology Research Centre of
Ontario, and from the Research Grants Committee of Hong Kong.

is pathological and to analyze the expected size of the resulting deterministic finite-state machines. We follow the first avenue here; we explore the second avenue in a companion paper [9] based on the preliminary results of Leslie [8].

This paper raises more questions than it solves since it deals with applied algorithms and applied data structuring. Nevertheless, we believe that our suggestions are of interest in that they may encourage others to tackle the remaining questions.

2 Subset Construction

We assume that readers are familiar with the notions of a finite-state machine; however, we review the basic definitions. A finite-state machine M is specified in a standard manner by a tuple $(Q, \Sigma, \delta, s, F)$, where Q is a finite set of states, Σ is a finite alphabet (the input alphabet), δ is a finite program given as a set of instructions of the form

$$(\text{source-state, source-symbol, target-state}),$$

s is the start state of M, and $F \subseteq Q$ is a set of final states. A computation of M on an input string $a_1 \cdots a_m$ corresponds to one executable sequence of instructions

$$(p_1, a_1, p_2)(p_2, a_2, p_3) \cdots (p_m, a_m, p_{m+1}),$$

where $p_1 = s$. If p_{m+1} is in F, then M accepts $a_1 \cdots a_m$. Given a finite-state machine M in which, for every state p and every symbol a, there is at most one instruction of the form $(p, a, ?)$ in δ, then M is deterministic. When M is nondeterministic, there may be many different M-computations on an input string.

It is not difficult to compile a deterministic finite-state machine into a subprogram of a given programming language and even into a VLSI chip. Compiling a nondeterministic finite-state machine is, however, more costly since the compiled code must incorporate backtracking to handle the nondeterminism. An elegant compilation was given by Thompson [12]; it has never been matched. An alternative and standard approach is to convert a nondeterministic finite-state machine into an equivalent deterministic finite-state machine using subset construction. (An intermediate approach is to do subset construction on the fly for those state sets reached by the input string. We do not discuss this idea any further in this paper.) Wood [13] gave the following algorithm for subset construction (clearly not intended for direct implementation):

Algorithm Subset Construction
On entry: A nondeterministic finite-state machine, $M = (Q, \Sigma, \delta, s, F)$.
On exit: A deterministic finite-state machine, $M' = (Q', \Sigma, \delta', s', F')$
    that satisfies $L(M) = L(M')$.
begin
    Let $Q_0 = \{s\}$ be the zeroth subset of Q and 0 be the
    corresponding state in Q', let i be 0, and let *last* be 0;

```
    while i ≤ last
    do begin
        for all a in Σ
        do if {(p, a, q) : p ∈ Q_i} ≠ Q_j, for some j, 0 ≤ j ≤ last
            then begin last := last + 1;
                δ' := δ' ∪ {(i, a, last)};
                Q_last := {q : p ∈ Q_i and (p, a, q) ∈ δ}
            end;
            i := i + 1
    end {while};
    Q' := {i : 0 ≤ i ≤ last};
    F' := {i : Q_i ∩ F ≠ ∅ and 0 ≤ i ≤ last}
end
```

Note that the core part of this algorithm is to take a reachable set Z of states that has not yet been examined and to produce, for each symbol a in Σ, the set T_a of states that M can reach with a single instruction that reads a. Formally, for all a in Σ,

$$T_a = \{t : p \in Z \text{ and } (p, a, t) \in \delta\}.$$

There are two varieties of subset construction: **big bang** and **reachable**. With big-bang subset construction, we begin with all $2^n - 1$ subsets of the states Q, where $n = \#Q$, and, for each subset and each symbol, we determine the state set reachable with one instruction. With reachable subset construction on the other hand, we carry out the instruction computation on only those state sets that are reachable from the state set $\{s\}$. Mathematically, these constructions are equivalent once we remove unreachable state sets after using the big-bang construction. Algorithmically, the reachable construction is more complex since we have to avoid duplicated work by considering only reachable state sets that we have not previously examined.

Thus, we have two tasks to implement efficiently: instruction computation and reachable-state-set maintenance. We focus on the task of instruction computation. In the companion paper [9], we address the issue of reachable-state-set maintenance.

3   Instruction Computation

A simple algorithm for instruction computation, given a state set Z, is as follows:

Algorithm Instruction Computation I
On entry: A nondeterministic finite-state machine, $M = (Q, \Sigma, \delta, s, F)$, and a state set Z.
On exit: The set $\{(Z, a, T_a) : a \in \Sigma\}$ of instructions.
begin
    for all $a \in \Sigma$
    do begin $T_a := \emptyset$;

```
        for all p ∈ Z
            do for all q ∈ Q
                do if (p, a, q) ∈ δ
                    then T_a := T_a ∪ {q}
    end;
    return {(Z, a, T_a) : a ∈ Σ}
end
```

It is not our aim to analyze this algorithm's execution time precisely; we just want to point out that it can perform badly.

Note that we have the Σ-loop as the outer loop to ensure that, for each a in Σ, we compute T_a completely before examining the next symbol in Σ. This algorithm takes $\Omega(\#\Sigma \cdot \#Z \cdot \#Q)$ time with a simple-minded representation and implementation. Indeed, the statement "**if** $(p, a, q) \in \delta$" takes $O(\#\delta)$ time if we use exhaustive search.

What execution time should be our goal? (Over the years, whenever we have discussed this issue with fellow researchers, they invariably suggested that subset construction should take time linear in the size of the output machine. Our analysis, as coarse as it is, suggests otherwise.) It seems that, at least, we have to examine each state p in Z and each instruction (p, a, q) in δ, where $a \in \Sigma$, to obtain the instruction (Z, a, T_a). Let δ_Z denote the restriction of δ to the state set Z; then, we should shoot for $O(\#Z + \#\delta_Z)$ time. Since $\#\delta_Z \leq \#Z \cdot \#\Sigma \cdot \#Q$, this bound is smaller than the true execution time of Algorithm I. The natural question is: *Can we do better?*

We could implement the instruction computation in a completely different way as follows:

Algorithm Instruction Computation II
On entry: A nondeterministic finite-state machine, $M = (Q, \Sigma, \delta, s, F)$, and a state set Z.
On exit: The set $\{(Z, a, T_a) : a \in \Sigma\}$ of instructions.
begin
    for all $a \in \Sigma$ **do** $T_a := \emptyset$;
    for all $(p, a, q) \in \delta$
        do if $p \in Z$ **then** $T_a := T_a \cup \{q\}$;
    return $\{(Z, a, T_a) : a \in \Sigma\}$
end

This version avoids the test of whether a specific triple (p, a, q) is in δ by using only triples that are instructions in M. The beauty of this algorithm is that it has one initialization loop that takes $O(\#\Sigma)$ time and one computation loop that takes $\Omega(\#\delta)$ time. The computation's efficiency depends on how fast we can perform the union operation, but a lower bound is $\Omega(\#\Sigma + \#\delta)$ time.

Since $\#\delta \leq \#\Sigma \cdot \#Q^2$, the run time could be excessive; however, Algorithm II suggests a more efficient implementation is possible.

In INR [6] and Grail [11], we represent δ as a sorted sequence of instructions, where we use the standard lexicographic order of tuples to sort them. In addition, we represent states and symbols by nonnegative integers; hence, we represent state sets as sorted sequences of integers. Thus, in a sorted δ, all instructions with the same source state and all instructions with the same source state and the same input symbol appear as contiguous sorted subsequences. In this setting, the goal of instruction computation is to compute, for a given sorted sequence Z of integers and for all symbols $a \in \Sigma$, the sorted sequences T_a and the corresponding instructions (Z, a, T_a). We begin by ignoring the problem of producing T_a as a sorted sequence. The next algorithm, Algorithm III, is specified as follows:

Algorithm Instruction Computation III
On entry: A nondeterministic finite-state machine, $M = (Q, \Sigma, \delta, s, F)$,
and a state set Z.
On exit: The set $\{(Z, a, T_a) : a \in \Sigma\}$ of instructions.
begin
    for all $a \in \Sigma$ **do** $T_a := \emptyset$;
    for all $p \in Z$
        do for all $(p, a, q) \in p$-block
            do $T_a := T_a \cup \{q\}$;
    return $\{(Z, a, T_a) : a \in \Sigma\}$
end

Observe that Algorithm III takes $\Omega(\#Z + \#\delta_Z)$ time; thus, we are approaching our goal. We next explore two different ways to organize this computation. The first method, Algorithm IV, is a variation of Algorithm III that makes use of an additional **instruction-block index** into the sorted δ. The idea is that the block index gives, for each source state, the location of the first instruction of its block of instructions. Thus, we can access the first instruction of an instruction block in constant time. Apart from this change, Algorithm IV is close in spirit to Algorithm III. In Grail, we do a binary search of the sorted instructions, rather than use a block index.

Algorithm Instruction Computation IV
On entry: A nondeterministic finite-state machine, $M = (Q, \Sigma, \delta, s, F)$,
and a state set Z.
On exit: The set $\{(Z, a, T_a) : a \in \Sigma\}$ of instructions.
begin
    for all $a \in \Sigma$
    do begin $T_a := \emptyset$;
        for all $p \in Z$
            do begin $ip := index[p]$;
                while $instruct(ip).symbol = a$
                do begin
                    $T_a := T_a \cup \{instruct(ip).target\}$;
                    $ip := ip + 1$
                end
        end

end;
  return $\{(Z, a, T_a) : a \in \Sigma\}$
end

The advantage of Algorithms III and IV is that they do not examine irrelevant instructions. Note that Algorithm IV processes all instructions with the same symbol at the same time. The advantage of this approach is that we need keep only one state-set variable T that we reuse for each symbol. We expect Algorithm IV to be faster than Algorithm III since it does not do any membership tests. The major problem raised by Algorithms III and IV is: How do we organize the computation of the target state set so that we obtain a sorted state set? Therefore, we still ask: *Can we do better?* We consider an alternative approach in Section 4.

4 Replacement–Selection

We propose the use of **replacement–selection**, a technique found in the standard method of implementing external sorting. In external sorting, we use a k-way merge of k sorted runs, as they are called. To implement a k-way merge we use a **tournament** [7] or **heap** [2, 7, 14] of size k such that each entry in the heap is the next item in one of the runs. We use a **minimum heap** in which the root item has the smallest value in the heap and is the next item in sorted order to be appended to the output run. We select the root item and replace it with the next entry in its run. Of course, the next entry may not be the smallest item in the heap; therefore, we need to "trickle it down" to an appropriate location [2, 3, 14]. The important idea is that the heap structure is fixed; we only select and replace items. If we have k runs with a total of N items, we can k-merge them in $O(N \log k)$ time.

How is replacement–selection related to instruction computation? Basically, the idea is that if $\#Z = k$, then we treat the k instruction blocks as k runs and perform a k-way merge.

We do not give precise algorithms as we have done in the preceding section; instead we briefly describe the possible approaches to using replacement–selection. The most elegant technique, if not the most efficient, is that we organize a heap using the last two components of an instruction as its priority. Thus, the root of a minimum heap will hold a couple (a, q) such that $(p, a, q) \in \delta_Z$ and such that all other couples in the heap are no smaller than it (no earlier in sorted order). When we select the root value, we replace it with the couple of the next instruction in its p-block. If the next couple has the form $(a, ?)$, it will be selected at some point during the computation of T_a. If, on the other hand, it has the form $(b, ?)$, for some $b > a$, then it will not be selected until the computation of T_a is completed. Notice that T_a will be output in sorted order as required. This algorithm takes $O(\#Z + \#\delta_{Z,a} \log \#Z)$ time to compute T_a, where $\delta_{Z,a}$ denotes the restriction of δ with Z and a.

The second approach is to treat replacement–selection conceptually and not to implement it with a heap. Essentially, we order the heap only on the symbol of the instructions; thus, we must destroy the sorted order of the target states. The advantage of this approach is that replacement is faster since most of the time there is no trickle down, since the next instruction from a p-block will have the same symbol as the current instruction if there is nondeterminism. Of course, the target-state sequence will not be sorted, so we have to sort it in a postprocessing phase. Now observe that we can remove the heap completely, since all we are really doing is extracting all instructions with a source state from Z and an input symbol a. If we use one of the standard sorting algorithms such as Quicksort, we can compute T_a in $O(\#Z + \#\delta_{Z,a} \log \#\delta_{Z,a})$ time.

The advantage of the second approach is that we no longer have to order the couples on the fly; instead, we have to sort the target states afterward. The second approach is faster if we can sort a set T_a rapidly. As we have a subset T_a of a finite set Q of states, bucket sort is applicable (and perhaps even radix sort). Thus, we can compute T_a in $O(\#Z + \#\delta_{Z,a})$ time in the expected case. Observe that this bound is the one that we were shooting for in the *worst case*. If we are able to use radix sort or bit-vector sort, then we can achieve the same time bound in the worst case. (Essentially, INR uses a bit-vector sort when the subset is large with respect to Q; it uses a standard sorting method otherwise.) Theoretically, this is the end of the story; practically, we are not finished. Can we use radix sort or bit vector sort? Are they truly fast in practice?

Lastly, we often have to do subset construction when a machine's program also has null-input instructions (often called ϵ or λ instructions). A null-input instruction does not read an input symbol; it has the form

$$(\text{source-state}, \lambda, \text{target-state}),$$

where λ denotes the null or empty string. We can transform a machine with null-input instructions into an equivalent machine without null-input instructions using a standard algorithm; see the texts of Hopcroft and Ullman [5] and Wood [13]. This construction produces a machine that has size at most quadratic in the size of the original machine. We can avoid the explicit construction of this intermediate machine by computing its instructions on the fly as needed. Essentially, if we are computing an instruction (Z, a, T_a), for a given state set Z and input symbol a, then, whenever there is a computation of the form

$$(p_1, \lambda, p_2) \cdots (p_{m-1}, \lambda, p_m)(p_m, a, p_{m+1}),$$

for a state p_1 in Z, we add p_{m+1} to T_a. We can precompute the λ-reachable states from each state of the machine, so that we do not need to compute the p_{m+1} on the fly. Can we attain the goal of $O(\#Z + \#\delta_Z)$ time, in the expected and worst cases, when we also have to handle null-input instructions?

References

1. A.V. Aho. Algorithms for finding patterns in strings. In J. van Leeuwen, editor, *Algorithms and Complexity*, volume A of *Handbook of Theoretical Computer Science*, pages 255–300. The MIT Press, Cambridge, MA, 1990.

2. A.V. Aho, J.E. Hopcroft, and J.D. Ullman. *Data Structures and Algorithms.* Addison-Wesley, Reading, MA, 1983.

3. Th. H. Cormen, C. E. Leiserson, and R. L. Rivest. *Introduction to Algorithms.* MIT Press, Cambridge, MA, 1990.

4. M. Crochemore and W. Rytter. *Text Algorithms.* Oxford University Press, Oxford, England, 1994.

5. J.E. Hopcroft and J.D. Ullman. *Introduction to Automata Theory, Languages, and Computation.* Addison-Wesley, Reading, MA, 2 edition, 1979.

6. J.H. Johnson. INR: A program for computing finite automata, 1986.

7. D.E. Knuth. *The Art of Computer Programming, Vol. 3: Sorting and Searching.* Addison-Wesley, Reading, MA, 1973.

8. T. K. S. Leslie. Efficient approaches to subset construction. Technical Report CS-92-29, Department of Computer Science, University of Waterloo, Waterloo, Ontario, Canada, 1992.

9. T. K. S. Leslie, D. R. Raymond, and D. Wood. The expected performance of subset construction, 1996.

10. D. Perrin. Finite automata. In J. van Leeuwen, editor, *Formal Models and Semantics*, volume B of *Handbook of Theoretical Computer Science*, pages 1–57. The MIT Press, Cambridge, MA, 1990.

11. D. R. Raymond and D. Wood. Grail: Engineering automata in C++, version 2.5. Technical Report HKUST-CS96-24, Department of Computer Science, Hong Kong University of Science & Technology, Clear Water Bay, Kowloon, Hong Kong, 1996.

12. K. Thompson. Regular expression search algorithm. *Communications of the ACM*, 11:419–422, 1968.

13. D. Wood. *Theory of Computation.* John Wiley & Sons, Inc., New York, NY, 1987.

14. D. Wood. *Data Structures, Algorithms, and Performance.* Addison-Wesley, Reading, MA, 1993.

Building Automaton
on Schemata and Acceptability Tables

Application to French Date Adverbials

Denis Maurel

LI/E3i/Université François Rabelais
64 avenue Jean Portalis, 37200 Tours, France
maurel@univ-tours.fr

Abstract. This paper presents a lexical finite states automaton to parse French Date Adverbials. To build this automaton, I have developed an original model of representation (Schemata and Acceptability Tables), the computation and use of which I will explain in this paper.

1. Motivation.

A large part of Natural Language Processing may be conveniently analyzed by means of finite-state automata. An automaton can be read very quickly, the description is flexible [3], and it can be represented in a very compact form [17]. Finite-state models can be used effectively on such diverse subjects as phonology [6][7], speech recognition [1], morphology [5][19], local grammar [12], and syntactic analysis [14][16][18].

The purpose of this paper is to present the development of a lexical automaton to parse French Date Adverbials, such as the following extracts from the French newspaper Le Monde (dated January 13rd, 1992):

... Le président... et une délégation... se sont mis d'accord, *mercredi 8 janvier* (*Wednesday, January 8th*)... après le coup d'Etat du *30 septembre dernier* (*last September 30th*).

[The president and a delegation have come into agreement on Wednesday, January 8th, after the coup of last September 30th.]

... M. Théodore est secrétaire général... *depuis 1978* (*since 1978*)....

[Mr. Théodore is secretary-general since 1978.]

... le calme régnait, *en fin de matinée lundi 13 janvier* (*at the end of the morning oj Monday, January 13th*), dans les rues...

[Quietness prevailed, at the end of the morning Monday, January 13th, in the streets...]

...Ce choix devrait faciliter le règlement d'une crise devenue *chaque jour* (*everyday*) plus dramatique... M. Théodore pourrait commencer à travailler *au cours des prochaines semaines* (*during the next few weeks*)...

[This choice would have made easier the settlement of a crisis which was becoming everyday more tragic. Mr. Theodore could have begun to work during the next few weeks.]

... Les dirigeants... se sont *pour le moment* (*for the moment*) contentés de dénoncer... ce bouleversement préparé *depuis des jours* (*for several days*). Un... organisme... devrait être créé *prochainement* (*soon*)...

[The leaders are content for the moment to denounce this upheaval, which has been prepared for several days. An established body would have to be created soon.]

... *A la même heure* (*At the same hour*), M... Juppé... a exprimé l'espoir que... *A la mi-journée* (*At the middle of the day*)... M... Soisson... *Tard dans la soirée* (*Late in the evening*)... M... Roseau...

[At the same hour, Mr. Juppé has expressed the hope that... At the middle of the day, Mr Soisson... Late in the evening, Mr. Rousseau...]

It is impossible to build this automaton on a list, because there are too many adverbials to describe and to custom directly. For this reason, this automaton will be developed using an original model of representation: Schemata and Acceptability Tables.

One automaton will be constructed for dates, and another for hours. The model and the development of a lexical automaton for time adverbials will be examined in detail, and schemata presented for other adverbials. Finally, this paper will include some computational results.

2. Automata for French dates and hours

The French dates automaton is presented in Fig. 1[1]; although the labels are words, but they also represent whole classes of word (i.e. jeudi [Thursday] also represents every other day of the week). This automaton recognizes dates, such as *le matin du*

---

[1] None of these automota have directed edges indicating direction of flow of information, because it is always from left to right. ε means empty transition.

jeudi 29 février 1996 [on Thursday morning, February 29th 1996] or *en février 1996* [on February 1996].

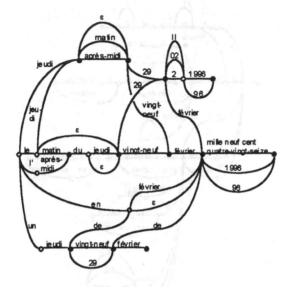

Fig. 1. French dates automaton

The automaton of Fig. 2 recognizes hours such as *à huit heures cinq* [at five past height] ou *à midi moins le quart* [at a quater to twelve].

3. Buiding automaton for French Time adverbs

The proposed model of representing French time adverbs such as *aujourd'hui* [today] or *demain* [tomorrow] is based on Harris' theory of sentence schemata and acceptability [4]. The set of patterns possible with time adverbs is presented by an automaton as a schema of time adverbs (Fig. 3), which will be able to recognize patterns such as *Preposition (Prép) Time adverb (Advd) Part of the day (Npj)* (e.g. *à partir de demain matin* [from tomorrow morning]).

Classes of word are also implemented by automata (Fig. 4).

But this automaton recognize also bad sequences like **aujourd'hui matin* [today morning]. So I note acceptability in binary matices, the acceptability tables. Here, we need two tables (Fig. 5 and 6) to correct our first description. At the intersection of each column and row a plus or minus sign indicates the acceptability or not of the corresponding combination.

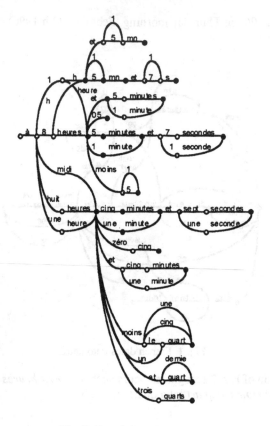

Fig. 2. French hours automaton

Fig. 3. French time adverbs schema

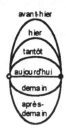

Fig. 4. *Npj* class automaton

	aujourd'hui	tantôt	hier	avant-hier	demain	après-demain
à	-	-	-	-	-	-
à partir de	+	+	+	+	+	+
après	+	+	+	+	-	-
aussitôt	-	-	-	-	-	-
avant	+	+	-	-	+	+
dans	-	-	-	-	-	-
depuis	+	+	+	+	+	+
dès	+	+	+	+	+	+
d'ici	+	+	-	-	+	+
d'ici à	+	+	-	-	+	+
durant	-	-	-	-	-	-
en	-	-	-	-	-	-
jusque	+	+	+	+	+	+
jusqu'à	+	+	+	+	+	+
jusqu'en	-	-	-	-	-	-
jusque vers	+	+	+	+	+	+
passé	+	+	+	+	+	+
pendant	-	-	-	-	-	-
pour	+	+	+	+	+	+
sous	-	-	-	-	-	-
sur	+	+	+	+	+	+
vers	+	+	+	+	+	+
au long de	-	-	-	-	-	-
au cours de	-	-	-	-	-	-
en cours de	-	-	-	-	-	-
avec	-	-	-	-	-	-
à travers	-	-	-	-	-	-

Fig. 5. The matrix *Prép_Advd*

	matin	midi	après-midi	tantôt	soir
aujourd'hui	-	+	-	+	-
tantôt	-	-	-	-	-
hier	+	+	+	+	+
avant-hier	+	+	+	+	+
demain	+	+	+	+	+
après-demain	+	+	+	+	+

Fig. 6. The matrix Advd_Npj

However, this automaton will also recognize bad sequences such as *aujourd'hui matin* [today morning]. To correct this, pattern acceptability has been charted into two binary matrices (acceptability tables) (Figures 5 and 6). At the intersection of each column and row, a plus or minus sign indicates whether or not the combination is acceptable. For example, *à partir de demain* [from tomorrow on] is acceptable, whereas **d'ici à hier* [from now to yesterday] is not (Fig. 5). In the same way, *demain matin* [tomorrow morning] is acceptable, whereas *aujourd'hui matin* [today morning] is not (Fig. 6).

All the minus signs of these two matrices were then incorporated into a deterministic minimal automaton of inacceptability (the local grammar A_2) (Fig. 7).

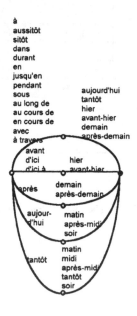

Fig. 7. Local grammar

This grammar was then applied to the deterministic minimal automaton of schema (A_1 - Fig. 4). Letting A be the desired result, L(A) to be the language recognized by A, and Λ the alphabet, we would like A to be such that:

$$L(A)=L(A_1)-\Lambda^*L(A_2)\Lambda^*$$

By implementing into A Mohri's algorithm of failure functions[15], we obtain the automaton of Fig. 8, which reads both automata (A_1 and A_2 with a failure function) and cuts transitions defining a word recognized by A_2.

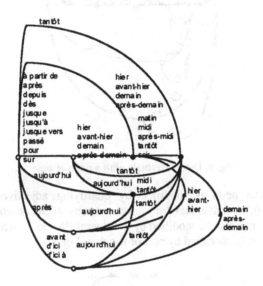

Fig. 8. Time adverbs automaton

4. French date adverbials schemata

Following the general outline of the automaton shown in Fig. 3, a number of schemata for French date adverbials have also been constructed. We will first demonstrate schemata developed for those adverbial phrases constructed using a noun of time (e.g. for three days), followed by those using a noun of date (e.g. since February 29), then by those using a noun of feast (e.g. after Christmas), and finally by those using presentative phrases (e.g. it is five years).

4.1. With a noun of time

Generally, adverbial phrases incorporating a noun of time are prepositional nominal phrases, consisting of a preposition, a determiner, and a noun of time (*Ntps*). The determiner is most commonly an article (*le, les*), a demonstrative (*ce, ces*), or a numeral (*Dnum*) (Fig. 9).

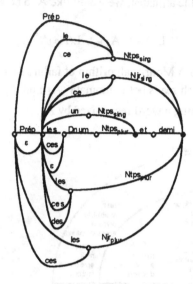

Fig. 9. Date adverbials with a noun of time

Occasionally the noun is modified by qualifying adjectives (*Adj*), ordinal adjectives (*Adjord*), or adverbs (*Adv*). Some nouns such as *été* [summer] and *décennie* [decade] may also be modified by a numeral (the date), while others such as *veille* [eve] (*Njr*) can be modified by part of the day (Fig. 10).

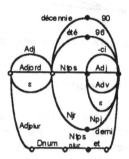

Fig. 10. Modifying noun of time

There are also more complex phrases [2] consisting of a predeterminer (*presque* [nearly] - *Pred*), an adjectival determiner (*chaque* [every] - *Dadj*), an adverbial determiner (*autant de* [as much] - *Dadv*), and a nominal determiner (*une moitié de* [half of] - *Dadj*) (Fig. 11).

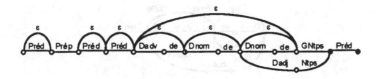

Fig. 11. Complex determiners

Locative determiners (*Dloc*) such as *début* [beginning], *milieu* [middle] and *mi-* [mid-] also present a problem (Fig. 12).

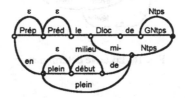

Fig. 12. Locative determiners

4.2. With a date

The same constructions exist for dates. In schemata, one distinguishes dates that begin with a day or month from those that distinguish hours. Four separate automata are presented below (Figures 13-16).

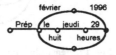

Fig. 13. Date phrases

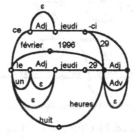

Fig. 14. Modifying date

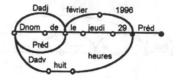

Fig. 15. Complex determiners

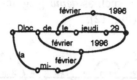

Fig. 16. Locative determiners

4.3. With a noun of feast

In the same way, there are here four automata to recognize date adverbials with a noun of feast. I have chosen days that are public hollidays in France (*GNfête*), like *Noël* [Christmas]. I add on these schemata the specific determiners of place *le pont de* [extra days off] (Figures 17-20).

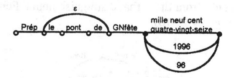

Fig. 17. Feast phrases

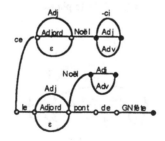

Fig. 18. Modifying noun of feast

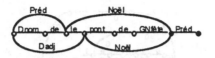

Fig. 19. Complex determiners

Fig. 20. Locative determiners

4.4. With presentative phrases

Le Bidois [8] names *presentative phrase* adverbials such as *il y a* and *voici* [it is]. Fig. 21 presents the schema for these adverbials.

Fig. 21. Presentative phrases

5. Computation

5.1. Dictionary

The dictionary of all words used in the class definitions is implemented using an h-transducer [17].

The initial pseudo-minimal automaton had 1 059 states and 1 431 transitions, with an alphabet of 56 letters. Minimization obtained an automaton with 878 states and 1 223 transitions.

The 499 words of this dictionary are coded between 0 and 498 (in the order of ASCII codes) by the h-transducer (ie. compacted into an array of 1 246 units). An example of such an h-transducer is given in Fig. 22, where the twelve months of the year are coded between 0 and 12.

For instance, the code of *mars* [March] is the addition of transition outputs:

$$m\,(7)\,a\,(+0)\,r\,(+1)\,s\,(+0) \rightarrow 8$$

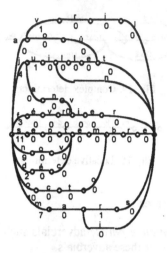

Fig. 22. A h-transducer for months

5.2. Acceptability tables

As was previously shown, schemata are corrected by acceptability tables. Fifty such tables have been constructed by me to describe French date adverbials [9], that is, 900 rows and 822 columns; space limitations prevent me from showing them all here.

To correct or to customize these tables, a finite-state transducer was used to assign a distributional class to every word (Fig. 23).

Both schemata and acceptability tables must be taken into account (Fig. 24).

5.3. Grammars

The automata of these schemata contain 132 states and 353 transitions for 135 classes. When these classes are replaced by sub-automata, 938 states and 8 826 transitions are obtained. These are transformed, by determinization, to 866 states and 7 954 transitions and, by minimization, to 514 states and 4 252 transitions. As previously noted, this automaton must be corrected using the constraints described in the acceptability tables, because not all schemata give rise to lexically correct temporal expressions.

Finally, the acceptability tables have been converted into an automaton with 542 states and 4 191 transitions. This automaton contains unacceptable local sequences, constituting a local grammar. After determinization, 1 488 states and 60 531 transitions exist, while after minimization, 603 states and 21 570 transitions remain.

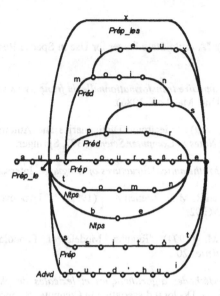

Fig. 23. A transducers for words begining by *au*

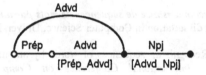

Fig. 24. Reading both schema and tables

5.4. Use

This automaton can be used to recognize French date adverbials [10], to constitute a module of presyntactic analysis [11], to compute dates [12] or to query a data base [13].

Acknowledgements

The author thanks anonymous reviewers for their comments. He thanks also Caroline Sori who has corrected a lot of english mistakes and Darell Raymond who has red this paper.

References

1. Church K. W. (1983): "A finite-State Parser for Use in Speech Recognition", *ACL* 83, the MIT Press.

2. Gross M. (1977): *Grammaire transformationnelle du français : syntaxe du nom*, Larousse, Paris (Reprint: Cantilène, Malakoff, 1986).

3. Gross M., Perrin D.(1989) : "Electronic Dictionnaries and Automata in Computationnal Linguistics", *Lecture Notes in Computer Science*, 377, Springer.

4. Harris Z. S. (1968) : *Mathematical Structures of Language*, Interscience Publishers.

5. Karttunen L., Kaplan R. M., Zaenen A. (1992): "Two-level Morphology with Composition", *COLING* 92.

6. Kaplan R. M., Kay M. (1994) "Regular Models of Phonological Rule systems", *Computational Linguistics* 20 .

7. Laporte E. (1988) : *Méthodes algorithmiques et lexicales de phonétisation de textes : applications au français*, Doctoral dissertation in Computer Science, University Paris 7.

8. Le Bidois G. (1935) : *Syntaxe du français moderne*, Paris, Picard.

9. Maurel D. (1989) : *Reconnaissance de séquences de mots par automate, adverbes de date du Français*, Doctoral dissertation in Computer Science, University Paris 7.

10. Maurel D. (1990) : "Recognizing sequences of words by automata ; the case of French adverbials of date", *International Conference on Computational Lexicography*, Balatonfüred, Hungary.

11. Maurel D. (1992) : "Preanalysis of French adverbials of date", 4th *European Workshop Semantics of Time, Space, and Movement and Spacio-temporal reasoning*, University of Toulouse, France.

12. Maurel D., Mohri M. (1994) : "French Temporal Expressions : Recognition, Parsing and Real Computation", *10 OED Conference*, Waterloo, Ontario, Canada.

13. Maurel D., Mohri M. (1995) : "Computation of French Temporal Expressions to query database", *NLDB'95*.

14. Mohri M. (1993) : *Analyse et représentation par automates de structures syntaxiques composées*, Doctoral dissertation in Computer Science, University Paris 7.

15. Mehryar Mohri, 1994. Compact Representations by Finite-State Transducers. In *Proceedings of the 32nd Annual Meeting of the Association for Computational Linguistics (ACL 94)*.

16. Pereira F., Wright R. N. (1991) : "Finite-State Approximation of Phrase Structure Grammars", Proceedings of the 29st Meeting of the *Association for Computational Linguistics*.

17. Revuz D. (1991): *Dictionnaires et lexiques - Méthodes et algorithmes*, Doctoral dissertation in Computer Science, University Paris 7.

18. Roche E. (1993): *Analyse syntaxique transformationnelle de français par transducteurs et lexique-grammaires*, Doctoral dissertation in Computer Science, University Paris 7.

19. Silberztein M. (1993): *Dictionnaires électroniques et analyse automatique de textes - Le système INTEX*, Paris, Masson.

FSA Utilities:
A Toolbox to Manipulate Finite-state Automata

Gertjan van Noord

Vakgroep Alfa-informatica & BCN
Rijksuniversiteit Groningen
vannoord@let.rug.nl

Abstract. This paper describes the *FSA Utilities* toolbox: a collection of utilities to manipulate finite-state automata and finite-state transducers. Manipulations include determinization (both for finite-state acceptors and finite-state transducers), minimization, composition, complementation, intersection, Kleene closure, etc. Furthermore, various visualization tools are available to browse finite-state automata. The toolbox is implemented in SICStus Prolog.

1 Introduction

This paper describes the *FSA Utilities* toolbox: a collection of utilities to manipulate finite-state automata and finite-state transducers. Manipulations include determinization (both for finite-state acceptors and finite-state transducers), minimization, composition, complementation, intersection, Kleene closure, etc. Furthermore, various visualization tools are available to browse finite-state automata. The toolbox is implemented in SICStus Prolog.

The motivation for the *FSA Utilities* toolbox has been the rapidly growing interest in finite-state techniques for *computational linguistics*. In particular, finite-state techniques are being used in computational phonology and morphology (Kaplan and Kay 1994), efficient dictionary lookup and part-of-speech tagging (Roche and Schabes 1995), natural-language parsing (Voutilainen and Tapanainen 1993), techniques for parsing ill-formed input (Lang 1989, van Noord 1995) and speech recognition (Oerder and Ney 1993, Pereira and Riley 1996), etc. The *FSA Utilities* toolbox has been developed to experiment with the techniques presented in these publications.

The following reasons for the popularity of finite-state techniques can be identified. Finite-state automata provide a well-studied, simple and very efficient formalism. Moreover, finite-state automata can be combined in various ways to construct larger automata from smaller ones. Such combined automata can still be processed very efficiently.

In this paper, I will illustrate the use of the *FSA Utilities* toolbox by means of a number of examples in section 3. I will then describe each of the operations which is provided by the *FSA Utilities* toolbox in more detail in section 4. Section 5 describes the visualization tools. Finally I will discuss some of the implementational issues in section 6. I first introduce the format for finite automata that is being used by *FSA Utilities*.

2 Representation

2.1 Finite-state Automata

A finite-state automaton is defined as a set of Prolog clauses. Some very limited familiarity with Prolog is assumed. A finite-state automaton is defined using the following relations:

start(State). State is the start state. There should be at least one start state.
    Multiple start states are supported.
final(State). State is a final state. There can be any number of final states.
trans(State0,Sym,State). There is a transition from State0 to State with
    associated symbol Sym. There can be any number of transitions.
jump(State0,State). There is an ϵ-transition from State0 to State. There
    can be any number of jumps.

It is assumed that states and symbols are ground Prolog terms (but see the discussion on constraints in section 2.3 below). A finite-state automaton defining the language consisting of any number of a's over the alphabet {a,b} is defined as follows:

$$\begin{array}{lll} \text{start(0).} & \text{final(0).} & \text{trans(0,a,0).} \\ \text{trans(0,b,1).} & \text{trans(1,a,1).} & \text{trans(1,b,1).} \end{array} \qquad (1)$$

It is worthwhile to note that we do not require that the transition relation is total. In other words, there can be states that have no outgoing transitions for certain symbols. In such cases we assume that these transitions do exist, but lead to some non-final state from which you cannot leave. This state is sometimes called the *sink*. In example 1 state 1 is such a sink. Thus, we may abbreviate example 1 as

$$\begin{array}{lll} \text{start(0).} & \text{final(0).} & \text{trans(0,a,0).} \\ \text{trans(0,b,1).} & & \end{array} \qquad (2)$$

In the representation we are using, we do not require that the alphabet and the set of states is defined explicitly. If the previous example is further abbreviated as:

$$\begin{array}{lll} \text{start(0).} & \text{final(0).} & \text{trans(0,a,0).} \end{array} \qquad (3)$$

the information that b is part of the alphabet is lost. This is problematic if the complement of this language is to be constructed; for this reason the difference operation has been introduced (cf. section 4.2).

2.2 Finite-state Transducers

Finite-state transducers are represented using the same conventions we use for finite-state acceptors, except that the symbol part of a transition now is written as a pair In/Out, indicating respectively the symbol that is read and the symbol that is written.

Moreover, the symbol '$E' is used to indicate ϵ, in order to allow transitions in which the input or output symbol is the empty symbol. This implies that there are two ways to define a jump in a finite-state transducer: trans(Q0,'$E'/'$E',Q) is equivalent to jump(Q0,Q). Note that the system does not allow the use of sequences of symbols in transitions (it can be shown that this does not restrict the transductions that can be defined).

A sub-sequential transducer is like a finite-state transducer except that we associate a sequence of symbols with a final state. These symbols are written to the output tape if the system halts in that final state. Such sub-sequential transducers are deterministic with respect to the 'in'-part of symbols: for each state there is at most one transition leaving that state upon reading some symbol. Note that we use sub-sequential transducers because the determinization of a finite-state transducer leads to a sub-sequential transducer (Roche and Schabes 1995). Sub-sequential transducers are represented as finite-state transducers, except that we now use a different predicate to indicate the final state with a sequence of symbols:

final_td(State,Symbols) indicates that State is a final state with associated sequence of symbols Symbols. Symbols is a ground Prolog list of symbols.

Sub-sequential transducers can be used wherever finite-state transducers are allowed.

2.3 Simple Prolog Constraints

Because finite-state automata are defined by Prolog clauses, it is very simple to attach Prolog constraints. For example, suppose that for a certain application it is useful to consider two subclasses of the alphabet, namely V (*vowels*) and C (*consonants*), we could then use the following technique to define the sequences $C^*V^*C^*$:

```
vowel(a).    vowel(e).    vowel(i).    vowel(o).    vowel(u).

cons(b).     cons(c).     cons(d).     cons(f).     cons(g).
cons(h).     cons(j).     cons(k).     cons(l).     cons(m).
cons(n).     cons(p).     cons(q).     cons(r).     cons(s).
cons(t).     cons(v).     cons(w).     cons(y).     cons(z).        (4)

start(0).    final(2).    jump(0,1).    jump(1,2).
trans(0,C,0) :- cons(C).
trans(1,V,1) :- vowel(V).
trans(2,C,2) :- cons(C).
```

I only allow constraints for which Prolog's built-in search procedure terminates. These Prolog constraints therefore do not increase the formal power of the formalism, although they are useful to define automata in a convenient and concise way.

3 Some examples

The *FSA Utilities* are available through a single UNIX command fsa which can take a number of different options. Suppose we have defined the following finite-state automaton, defining the language consisting of all the strings made up of an even number of a's followed by an even number of b's, in a file called aabb.nd:

```
start(0).          trans(0,a,1).        trans(1,a,0).
jump(0,2).         trans(2,b,3).        trans(3,b,2).      (5)
final(2).
```

In order to determinize this automaton, we give the following UNIX command (note that in these examples lines starting with a % are typed to the UNIX shell; the lines that follow are output of the *FSA Utilities* program):

```
% fsa -d <aabb.nd >aabb.d                                  (6)
```

This command writes the determinized automaton to the file aabb.d. This file now contains:

```
start(q0).         final(q0).           final(q1).
trans(q0,b,q2).    trans(q0,a,q3).      trans(q2,b,q1).    (7)
trans(q1,b,q2).    trans(q3,a,q0).
```

It is also possible to create an automaton on the basis of a regular expression. The file aabb.d could also be obtained by the command:

```
fsa -r '(a.a)* .(b.b)*' | fsa -d > aabb.d                  (8)
```

In such regular expressions the dot (.) is used to indicate concatenation, union is defined by the semi-colon (;) and Kleene closure is defined by the asterix (*).

A minimized automaton is obtained with the -m option. We can, for example, pipe the result of determinization to another incarnation of the fsa command. The following pipe produces the minimal finite-state acceptor for the language consisting of an even number of a's followed by an uneven number of b's:

```
% fsa -r '(a.a)* .(b.b)* .b' | fsa -d | fsa -m
start(q0).         final(q1).
trans(q0,a,q2).    trans(q0,b,q1).      trans(q1,b,q3).    (9)
trans(q2,a,q0).    trans(q3,b,q1).
```

Next consider a finite-state *transducer* which copies its input (strings of a's and b's) to its output, except that if an a is followed by a b, then this a becomes a b. Suppose this transducer is defined in the file a2b.tnd as follows:

```
start(0).              trans(0,b/b,0).        trans(0,a/b,1).
trans(1,b/b,0).        trans(0,a/a,2).        trans(2,a/b,1).     (10)
trans(2,a/a,2).        final(0).              final(2).
```

Such a transducer can be determinized (using the algorithm described in Roche and Schabes (1995) with the command:

```
% fsa -td <a2b.tnd >a2b.td                                        (11)
```

The file a2b.td now contains:

```
start(q0).             final_td(q0,[]).       final_td(q1,[a]).
trans(q0,b/b,q0).      trans(q0,a/'$E',q1).  trans(q1,b/b,q2).   (12)
trans(q2,'$E'/b,q0).   trans(q1,a/a,q1).
```

In order to use this transducer to transduce strings, it may be worthwhile to compile it into an efficient Prolog program implementing the transduction. The Prolog program will be fully deterministic; using first argument indexing, this determinism is visible to modern Prolog compilers. This implies that transduction is computed in linear time (with respect to the size of the input), and is independent of the size of the transducer.

```
% fsa -ct <a2b.td >a2b.pl                                         (13)
```

The file a2b.pl contains:

```
t_accepts(S,T) :- t_accepts_q0(S,T).
t_accepts_q0([],[]).
t_accepts_q1([],[a]).
t_accepts_q0([S|T],O) :- t_accepts_q0(S,T,O).
t_accepts_q1([S|T],O) :- t_accepts_q1(S,T,O).
t_accepts_q2([S|T],O) :- t_accepts_q2(S,T,O).                     (14)
t_accepts_q0(b,T,[b|O]) :- t_accepts_q0(T,O).
t_accepts_q0(a,T,O) :- t_accepts_q1(T,O).
t_accepts_q1(b,T,[b|O]) :- t_accepts_q2(T,O).
t_accepts_q1(a,T,[a|O]) :- t_accepts_q1(T,O).
t_accepts_q2(S,[b|O]) :- t_accepts_q0(S,O).
```

This Prolog program can be used to transduce the string aabbc by issuing the command:

```
% fsa -transduce a a b b < a2b.pl
a b b b                                                           (15)
```

In regular expressions it is possible to use pairs of symbols in order to obtain finite-state transducers. Thus we can have examples such as the following:

```
% fsa -r '((a:a;b:b;c:c)* .(d:e)* .(a:a;b:b;c:c)*)*' \
    | fsa -td
start(q0).             final_td(q0,[]).       trans(q0,d/e,q0).   (16)
trans(q0,c/c,q0).      trans(q0,b/b,q0).       trans(q0,a/a,q0).
```

As another example, it is possible to compose aabb.nd with the transducer a2b.tnd. The following pipe produces the minimized, determinized composition:

```
fsa -compose_fsa aabb.nd a2b.tnd | fsa -d \
             | fsa -m > result.d
```
(17)

Such automata can be inspected with the Tk Widget (presented in section 5), for example, as in figure 1.

In this paper the use of the *FSA Utilities* toolbox is illustrated by means of UNIX commands. However, this is not the only possible way of using the toolbox. Care has been taken to implement the toolbox in such a way that it is straightforward to use each of the operations as a Prolog library. For this reason most of the operations are defined in separate modules.

4 Operations on Finite Automata

The FSA Utilities toolbox provides a number of operations on finite-state automata. These operations are presented in this section. First I present the operations for finite-state acceptors, then I define the operations related to finite-state transducers.

4.1 Finite-state Acceptors

The -accepts Words option can be used to check whether a given string of symbols (Words) is accepted by a given finite state automaton (read from standard input). This automaton is defined according to the conventions discussed in section 2. Alternatively this file is a compiled finite-state automaton as produced by the -compile option presented below. The program determines whether Words is accepted by this automaton or not. If it is, the program returns successfully; otherwise it exits with exit code 1. Note that this procedure is only guaranteed to be linear in the length of the input string if the finite state automaton is both deterministic and compiled. In the case of such a compiled deterministic automaton SICStus Prolog's first argument indexing is exploited to ensure that acceptance is also independent of the size of the automaton.

The -produce option is used to generate all possible strings accepted by the input automaton. Strings are produced in increasing length. Clearly this operation need not terminate if the automaton defines a language consisting of an unbounded number of strings.

Consider the following examples.

```
% fsa -a a a a a b b < aabb.nd
yes
% fsa -a a a a b b < aabb.nd
no
% fsa -d <aabb.nd | fsa -c > aabb.pl
% fsa -a a a a a b b b b b b < aabb.pl
yes
```
(18)

```
% fsa -p <aabb.nd

a a
b b
a a a a
a a b b
b b b b
a a a a a a
a a a a b b
a a b b b b
b b b b b b
...
```
(19)

4.2 Regular expressions

The following options can be used to construct automata for regular languages
on the basis of the language(s) defined by one or more input automata:

```
-concat File1 File2
-kleene File1
-union File1 File2
-reverse File1
-intersect File1 File2
-complement File1
-difference File1 File2
```

These options can be used to combine automata by concatenation, Kleene closure, union, reversal, intersection, complementation or difference respectively.
Examples:

```
% fsa -complement <aabb.d
start(q4).           final(q5).           final(q6).
final(q7).           trans(q5,a,q5).      trans(q5,b,q5).
trans(q4,a,q7).      trans(q4,b,q6).      trans(q8,a,q5).
trans(q8,b,q6).      trans(q6,a,q5).      trans(q6,b,q8).
trans(q7,a,q4).      trans(q7,b,q5).
```
(20)

Because the system assumes that the alphabet consists only of the symbols
occurring in the input automaton, it may be safer to use the -difference option.
In this case an automaton is written to standard output, defining all strings that
are defined by the automaton in File1 except for the strings defined in File2. For
example, if the file abc.nd consists of the automaton defining all strings over
the alphabet {a,b,c}, then we get:

```
% fsa -difference abc.nd aabb.nd
start(q9).            final(q10).           final(q11).
final(q12).           trans(q9,a,q12).      trans(q9,b,q11).
trans(q9,c,q10).      trans(q10,a,q10).     trans(q10,b,q10).
trans(q10,c,q10).     trans(q11,a,q10).     trans(q11,b,q13).
trans(q11,c,q10).     trans(q12,a,q9).      trans(q12,b,q10).
trans(q12,c,q10).     trans(q13,a,q10).     trans(q13,b,q11).
trans(q13,c,q10).
```
(21)

The system also supports the construction of an automaton on the basis of a regular expression: the option **-r Expression** can be used. The operators that are allowed are Kleene closure (*), concatenation (.) and union (;). Moreover, it is possible to have transduction pairs (:). Finally, the system understands intervals. The expression **a-z** is a shorthand for the expression **a;b;c;..;z**. For example:

```
% fsa -r '(a*;b*).(d-g)*' | fsa -ps
```
(22)

produces the following postscript picture:

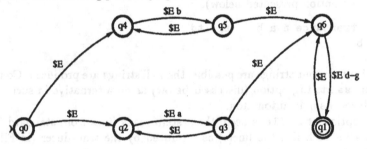

4.3 Determinization

The -d option is used to determinize a finite state automaton. Standard input consists of the finite-state automaton to be determinized. The determinized finite-state automaton is written to standard output. The determinization algorithm uses a version of the subset construction (Hopcroft and Ullman 1979) in which only those subsets are considered that can be reached from an initial state. Example:

```
% fsa -d aabbaabb.nd | fsa -tex -q 0 > aabbaabb.pic
```
(23)

This results in the following LaTeX picture of the determinized automaton:

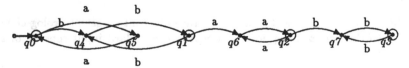

4.4 Minimization

Currently, the toolbox supports three different implementations of the minimization operation. The option -m indicates that the algorithm of Hopcroft and Ullman (1979) should be used. The input automaton should already be deterministic. The option -m2 indicates the use of Brzozowski's method (Brzozowski 1962). The input automaton does not need to be deterministic. The option -m3 indicates that Hopcroft's algorithm should be used (Hopcroft 1971). The input automaton must be deterministic. The implementation of these algorithms is described in section 6.3.

```
% fsa -d < y.nd | fsa -m > y.m
% fsa -m2 < y.nd > y.m
```
(24)

4.5 Finite-state Transducers

If the option -transduce Words is used, the system produces all possible transductions of Words on the basis of the finite-state transducer read from standard input. This transducer can be specified in compiled format (a .pl file resulting from the -ct option presented below).

```
% fsa -transduce a a b b <a2b.td
  a b b b
```
(25)

If multiple output strings are possible, then all strings are produced. Consider the -compose_string option (discussed below) as an alternative in such cases: this produces a single automaton.

If the option -ta File is used, the system reads lines from standard input and writes for each line the line(s) as produced by the transducer (in File) to standard output.

If the option -tp is used, a finite state transducer is read from standard input. This transducer need not be deterministic. All pairs of sentences in the mapping defined by the transducer are written to standard output. Pairs are produced in increasing length of the input string.

```
% fsa -td <a2b.a2b.tnd |fsa -tp
==>                          b ==> b
a ==> a                      b b ==> b b
b a ==> b a                  a b ==> b b
a a ==> a a                  b b b ==> b b b
b b a ==> b b a              b a b ==> b b b       (26)
b a a ==> b a a              a b b ==> b b b
a b a ==> b b a              a a b ==> b b b
...
```

The -ct option is used to compile a transducer into a set of Prolog clauses (similar to the -compile option for acceptors). Standard input consists of the finite-state automaton to be compiled. The compiled clauses are written to standard output. If the input is a deterministic transducer, then the compiled clauses can be used by Prolog deterministically for transduction (clauses are indexed such that the Prolog compiler is able to recognize that the application of each clause is deterministic); transduction time is then independent of the size of the automaton. Examples:

```
% fsa -ct <a2b.td >a2b.pl
```
(27)

The options -domain and -range can be used to obtain the domain (range) of the mapping defined by the transducer read from standard input. The -identity option can be used to obtain a transducer defining the identity relation for the language defined by the finite-state automaton read from standard input:

```
% fsa -domain <a2b.a2b.tnd |fsa -d |fsa -m
start(q0).          final(q0).          trans(q0,a,q0).
trans(q0,b,q0).
% fsa -identity <aabb.nd                                      (28)
start(q0).          final(q1).          trans(q0,a/a,q2).
trans(q2,a/a,q0).   trans(q1,b/b,q3).   trans(q3,b/b,q1).
jump(q0,q1).
```

4.6 Composition

Transducers can be composed with the option -compose File1 File2. In this form, the two files are assumed to contain finite state transducers. The (serial) composition is written to standard output. In the form -compose_fsa File1 File2, the first file contains a finite state automaton and the second file contains a finite state transducer. The composition is written to standard output. In the form -compose_string Words, a finite-state transducer is read from standard input and composed with the (finite-state automaton defining the language consisting of the single) string Words. The composition is written to standard output. Examples:

```
% fsa -compose a2b.tnd a2b.tnd >a2b.a2b.tnd
% fsa -compose_fsa aabb.nd a2b.tnd
% fsa -compose_string a a b <a2b.tnd |fsa -d |fsa -m          (29)
start(q0).          final(q1).          trans(q0,a,q2).
trans(q2,b,q3).     trans(q3,b,q1).
```

Note that the compose_string operation differs from the transduce operation discussed previously. If the transducer is non-deterministic and relates Words to several output strings, then this option constructs a finite-state automaton defining exactly the set of those output strings; if the transduce option is used, then all of the resulting strings are written to standard output.

In computational morphology a different kind of composition is sometimes identified: parallel composition. The parallel composition of two transducers is obtained by the -intersect option introduced above for finite state acceptors.

4.7 Determinization of Transducers

Determinization of a transducer can be performed with the -td option. Standard input consists of the transducer to be determinized. The determinized transducer is written to standard output.

```
% fsa -td <a2b.a2b.tnd >a2b.a2b.td                                    (30)
```

The determinization algorithm for transducers is described in Roche and Schabes (1995). Not every transducer can be determinized. This procedure is guaranteed to terminate only if the input transducer *can* be determinized. Furthermore, note that the output transducer is a *sub-sequential* transducer: each final state is associated with a sequence of symbols that are outputted if the system halts in that state.

5 Visualization of Automata

The system offers a number of different means of visualization. Representations of finite-state automata can be constructed which are compatible with either the VCG (Sander 1995) or the daVinci (Fröhlich and Werner 1994) visualization programs. These programs are generic graph-visualization programs; the results for finite-state automata tend to differ a lot from example to example. VCG has no problems with larger graphs; daVinci sometimes has. Various options can be used to obtain different effects.

Furthermore, the system provides its own visualization through a Tk-widget (Ousterhout 1994). This widget allows dragging of states to manipulate the visualization of an automaton interactively. Moreover, LATEX picture output or Postscript output can be generated on the basis of this (interactively manipulated) visualization. The system can also output LATEX picture output or Postscript output directly, using the -tex and -ps options.

5.1 VCG

If the option -vcg is used, a representation compatible with the VCG 1.30 graph-visualization program is written to standard output, on the basis of the finite-state automaton read in from standard input. For example, if the file fig19.tnd consists of the transducer given in figure 19 of Roche and Schabes (1995), then the following pipe produces the illustration given in figure 2.

```
% fsa -td fig19.tnd | fsa -vcg | xvcg -                               (31)
```

5.2 DaVinci

If the -davinci option is used, a term-representation compatible with the requirements of the daVinci 2.0 graph-visualization program is written to standard output, on the basis of the finite-state automaton read in from standard input. An example of daVinci is given in figure 3.

5.3 Latex Picture

The -tex option is used to produce a LaTeX picture. The same visualization mechanism is used as for the -tk option, cf. below. Also, the same options to influence this visualization can be given.

% fsa -tex <ba_n.nd >ba_n.pic   (32)

5.4 Postscript

The option -ps is similar to the -tex option. In this case the output is given as a Postscript picture.

5.5 The Tk Widget

The option -tk [[-q Integer] [-xd Dist] [-angle Angle]] File indicates that a finite state automaton is read from File and shown in a Tk Widget. The -xd option can be used to alter the default X distance of nodes (default: 120). The -angle option indicates the angle of edges (default: 0.25). Straight lines can be obtained with a value of 0. The quality option indicates how hard the system should try to obtain a good visualization. Default is 1, which indicates that not much effort is made at reducing the number of crossing branches; larger integers will (dramatically) increase processing time, resulting sometimes in slightly better output.

The Tk Widget offers a graphical user interface to browse finite-state automata. For an illustration, consider figure 1. The use of this widget is explained now as follows.

The Canvas The canvas displays the finite-state automaton. The display can be controlled by a number of options to the fsa command, or on-line by using the button-bar.

By default, edges are not drawn as straight lines, but as curves. The angle of the edge is determined by the -angle option or the Edge Angle field on the button-bar.

The default distance between nodes is determined by the -xd option or the X-distance field on the button-bar.

Finally the system provides limited functionality to compute the most suitable geometry. The intent is to minimize the number of crossing edges, as well as the size of the area needed to display an automaton. Since in general such a procedure is much too costly, the -q option or the Quality field on the button-bar can be used to tell the system how much effort should be spent on this. A higher value should imply better quality.

The finite-state automaton can be manipulated manually by dragging states of the automaton to a new place. This possibility implies that you can adapt the visualization of the automaton to your own taste, and to the characteristics of a particular automaton.

The Button-bar

Save The current geometry of the finite-state automaton is saved in the file from which the current automaton was read. Note that the next time this file is used as input for the **fsa -tk** command, all geometry options are ignored, and this information is used instead.

SaveAs As the above, but the user is prompted for a file-name.

SaveAsTeX The user is prompted for a file-name. In this file the system outputs the current geometry of the automaton as a LaTeX picture. This can then be used in a LaTeX file, to give e.g.:

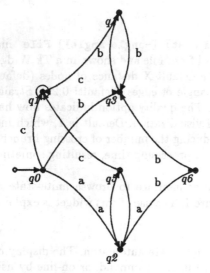

SaveAsPs The user is prompted for a file-name. In this file the system outputs the current geometry of the automaton as a Postscript picture.

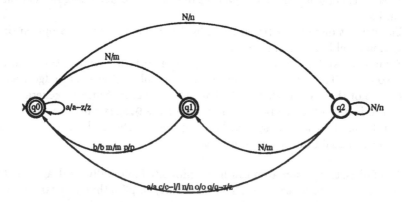

Load You are prompted for a file-name. The file-name is assumed to contain a finite-state automaton which will be displayed in the canvas.

Redraw The current automaton is redrawn.

Revert All changes are discarded and the automaton is redrawn.

Pause The Tk-Widget is removed from the screen.

Debugging When this option is set, progress information during the computation of the geometry is shown.

Quit The application is halted.

Edge Angle Determines the edge of an angle. When this is set to 0 then straight lines are used. Useful values are between 0 and 1.

X-distance Determines the distance between nodes.

Quality Determines the effort that should be spent on the computation of the geometry. If this value is larger than 0, the system tries to minimize the number of crossing edges and the size of the area needed for the automaton. Larger values dramatically increase the time needed to compute this.

Help The man page is displayed in an **xterm**.

6 A note on Implementation

Prolog is a high-level language in which operations on finite-state automata can be implemented rather straightforwardly.

However, Prolog lacks the appropriate data-structures to implement e.g. the subset-construction algorithm as efficiently as possible. For this reason I have developed a library which implements a data-structure which approximates a hash-table: this data-structure provides *log* time access rather than constant time access.

Furthermore, note that Prolog's built-in search method (backtracking) is not used in the implementation of the determinization and minimization operations. These operations are implemented by deterministic algorithms. If we are careful in designing the Prolog code for these algorithms, then the resulting Prolog code can be run efficiently, because modern Prolog compilers are capable of recognising determinism.

6.1 Determinization

The determinization algorithm I used is a subset-construction algorithm. For a given automaton (the 'input' automaton) the construction computes an equivalent, but deterministic automaton (the 'output' automaton). The algorithm maintains an agenda of states (of the output automaton) whose successors need to be computed, a table of states (of the output automaton) that have already been processed, a list of transitions (of the output automaton), and a list of final states (of the output automaton). A state of the output automaton is a subset of states from the input automaton. For each state on the agenda, the procedure computes all transitions that are possible leaving that state. If new states arise, then these states are added to the agenda.

The relation `treat_start_states` computes the set of states reachable without input consumption from the start states of the input automaton. This set of states from the input automaton constitutes the single start state of the output automaton. This start state is added to the list of final states if one of the input states is a final state. The initial agenda for the subset construction process contains this start state as a single element.

```
determinize([Start],Finals,Transitions) :-
    treat_start_states(Start,Table,Finals0,Finals),          (33)
    closure([Start],Table,Finals0,Transitions).
```

The relation `closure` recursively processes the agenda. Its arguments are respectively the list of states to be processed, the table of states already processed, the resulting list of final states, and the resulting list of transitions. If the agenda is empty, then the process finishes. Otherwise the first state of the agenda is considered and all transitions leaving this state are computed. Any new states that result are added to the agenda.

```
closure([],_table,[],[]). % no more states, finished.
closure([H|Agenda0],Table0,Tr,Fin) :-
    findall(NTrans,new_transition(H,NTrans),Transs),         (34)
    add(Transs,Agenda0,Agenda,Table0,Table,Tr0,Tr,Fin0,Fin),
    closure(Agenda,Table,Tr0,Fin0).
```

The relation `new_transition` defines what a possible transition is. The relation consists of a state (from the output automaton) and all transitions that are possible with this state as a source state.

```
new_transition(Begins,trans(Begins,Sym,Ends)) :-
    setof(End,a_trans(Begins,Sym,End),Ends).                 (35)

a_trans(Begins,Sym,End) :-
    member(Begin,Begins),
    trans(Begin,Sym,Mid),                                    (36)
    p_tr_jump(Mid,End).   % End is jump-reachable from Mid
```

The `p_tr_jump` relation consists of pairs of states from the input automaton such that the second state can be reached without reading any symbols from the first state (reflexive and transitive closure of the jump relation).

The relation `add/9` inspects the given list of transitions for new target states. These transitions are added to the list of transitions. Each new state (i.e. not yet in the table) is added to the agenda and to the table (and to the list of final states if this state contains a final state of the input automaton). This concludes the presentation of the determinization algorithm.

6.2 Determinization of Transducers

The determinization algorithm that is used for transducers is a generalization of the algorithm presented. It is based on the algorithm described in Roche and Schabes (1995). In this generalized version, a state in the determinized transducer is a set of states from the input transducer, where each of these states is associated with a sequence of symbols that still need to be written.

The output of this algorithm is a sub-sequential transducer. We can think of such a transducer as an ordinary transducer, except that a sequence of symbols is associated with each final state. If the system halts in a final state, then the associated symbols are written to the output tape. Such transducers are able to recognize whether or not the end of the string has been reached. Consider, for example, the following transducer. This transducer transduces a string of a's into b's, except for the last a:

```
start(q0).          final(q1).
trans(q0,a/b,q2).   trans(q2,a/a,q1).        (37)
jump(q0,q2).        jump(q2,q0).
```

In order to determinize this transducer, (i.e. ensure that for each state and each symbol there is only a single transition), the extra power of sub-sequential transducers is necessary:

```
start(q0).              final_td(q1,[a]).
trans(q0,a/'$E',q1).    trans(q1,a/b,q1).        (38)
```

Even if the input transducer defines a function, it is not always possible to construct such a deterministic (subsequential) transducer. An example of an inherently non-deterministic transducer is given below:

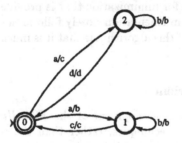

If the system is in state 0 and it reads an a, then it can only decide which transition to take after reading an unbounded number of b's. The algorithm runs into an infinite loop for transducers for which no determinized version exists. [1]

6.3 Minimization

The *FSA Utilities* toolbox provides implementations of three different minimization algorithms. For a comparison of these and other minimization algorithms, refer to Watson (1995).

---

[1] Note that the results discussed in paragraph IV.6 of Berstel (1979), based on (Choffrut 1977), suggest that it is decidable whether an input transducer can be determinized.

The first minimization algorithm is taken from Hopcroft and Ullman (1979). In our implementation the following phases can be identified. Firstly, transitions are added to the incoming automaton to ensure that the transition function is total (note that it is assumed that the input is already deterministic). Then, in the second step, all pairs of non-identical states are computed. In the third phase, the automaton is adapted in such a way that states that were found to be identical receive a single name. Finally, some unneccessary transitions are thrown away (those ending and starting in the 'sink' state).

The computation of the set of pairs of non-identical states starts out with pairs of states for which one element is a final state and the other element is not a final state: these clearly are non-identical. These pairs are taken as an agenda and we then proceed to find more pairs of non-identical states on the basis of these pairs. Such new non-identical pairs can be obtained by the following reasoning. If P and Q are non-identical states, and there are transitions trans(P1,Sym,P) and trans(Q1,Sym,Q) then P1, Q1 is a pair of non-identical states too.

The *FSA Utilities* toolbox also provides an implementation of Brzozowski's method of minimizing an automaton (Brzozowski 1962, Watson 1995). In this method the (possibly non-deterministic) automaton is reversed, determinized, reversed and determinized. Note that this method is often much faster than the previous method. Moreover the input can be non-deterministic. Finally, the implementation is very simple, since it consists only of automaton reversal (a particularly simple operation) and determinization (which is useful independently).

The third algorithm for minimization that is provided is the algorithm from Hopcroft (1971). Our implementation closely follows the presentation of Watson 1995. A disadvantage of this algorithm is that it is much more complicated than the other two.

6.4 Practical Experience

The main purpose of the development of the *FSA Utilities* toolbox has been to experiment with new techniques, rather than the construction of practical programs. However, one might still be interested in the practical efficiency of some of the operations provided. The determinization and minimization operations are the most critical operations. Minimization typically is much slower than determinization. The second version of the minimizer (using reversal and determinization twice) is much faster than the other versions. The current implementation of the determinizer runs much faster than a previous version, in which the transitive closure of jumps was computed as a separate pre-processing step. We also experienced that the second version of the minimizer runs much faster on a determinized automaton.

To get at least some idea of the system's practical behavior, first consider a finite-state automaton recognizing all mono-syllabic Dutch words. This non-deterministic automaton (written by Gosse Bouma) has 185 transitions, 9 jumps, 42 states and 26 symbols. Determinization takes 200 milliseconds (on a 1995 HP-UX 9000/735), resulting in an automaton with 49 states and 392 transitions. Minimization is much slower: it takes about 24 seconds to reduce the number of states to 36 and the number of transitions to 315. If Brzozowski's method is used, minimization can be done in less than three seconds and has the additional advantage that the input need not be determinized first. Hopcroft's method takes about 11 seconds.

As another example consider a finite-state *transducer* which translates temporal expressions such as *two minutes before half past seven* into 7 28. This automaton is larger: it has 5824 states, 8809 transitions, 67 input symbols and 61 output symbols. Determinization takes less than 5 seconds. The result has 1714 states and 3362 transitions. In comparison, in order to determinize the automaton defining the domain of the transduction (obtained by simply removing the output part of the symbols) takes less than three seconds. The resulting automaton can be minimized (using Brzozowski's method) in about four seconds.

Availability

The *FSA Utilities* toolbox is available under GNU General Public License from anonymous ftp. Alternatively you can obtain the toolbox via the World-wide Web. Addresses:

```
ftp://ftp.let.rug.nl/pub/prolog-app/FSA/
http://www.let.rug.nl/~vannoord/Fsa/
```

Acknowledgements

Feedback from Gosse Bouma, Mark-Jan Nederhof, John Nerbonne, Fernando Pereira, Edwin Kuipers and anonymous reviewers from the First International Workshop on Implementing Automata is gratefully acknowledged. The option to produce postscript output is based on work by Peter Kleiweg.

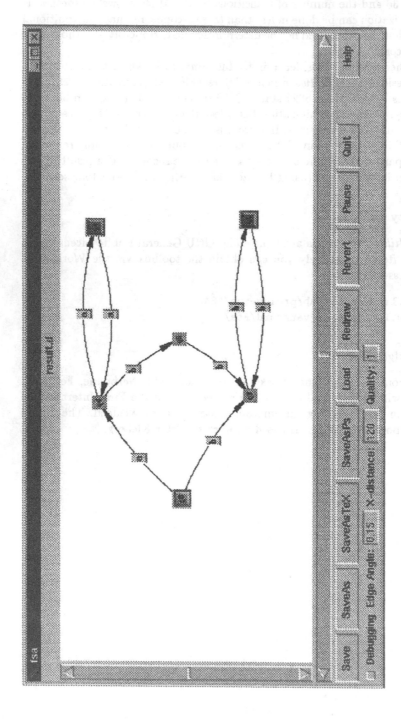

Fig. 1. The *FSA Utilities* toolbox uses Tcl/Tk for its built-in visualization of finite-state automata. The user can interactively alter the visualization by dragging states of the automaton to different positions. LaTeX picture output and Postscript output can be generated on the basis of the current view.

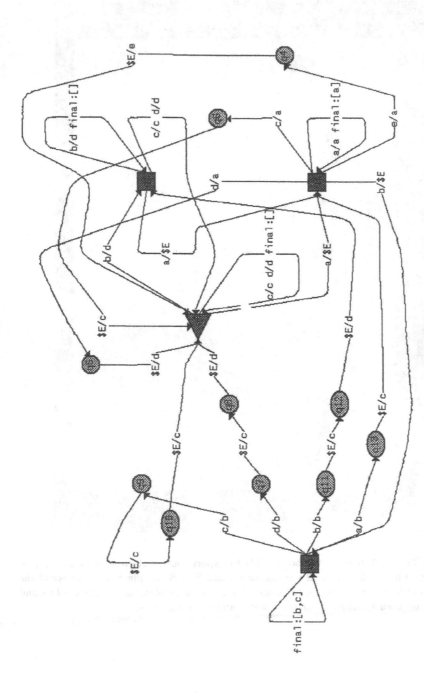

Fig. 2. The *FSA Utilities* program supports output for two external graph visualization programs. This figure illustrates the use of VCG. This example is the determinized version of an example given by Roche and Schabes (their figure 19).

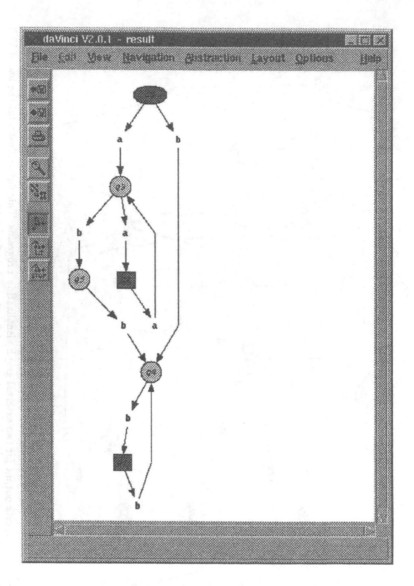

Fig. 3. The *FSA Utilities* toolbox is able to export finite-state automata in a representation suitable for external visualization tools. This figure illustrates the use of the daVinci visualization program to inspect a finite state automaton. Different colors and shapes are used to indicate the start states and the final states.

References

Berstel, Jean 1979. *Transductions and Context-Free Languages.* Teubner Studienbücher, Stuttgart.

Brzozowski, J.A. 1962. Canonical regular expressions and minimal state graphs for definite events. In *Mathematical theory of Automata.* Polytechnic Press, Polytechnic Institute of Brooklyn, N.Y., pages 529–561. Volume 12 of MRI Symposia Series.

Choffrut, Ch. 1977 Une caractérisation des fonctions séquentielles et des fonctions sous-séquentielles en tant que relations rationelles. *Theoretical Computer Science* 5, 325-338.

Fröhlich, M. and M. Werner. 1994. The graph visualization system *davinci* - a user interface for applications. Technical Report 5/94, Department of Computer Science; University of Bremen. Available by anonymous ftp: ftp://ftp.Uni-Bremen.DE/pub/graphics/daVinci/papers/techrep0594.ps.gz; cf. also http://www.informatik.uni-bremen.de/~inform/forschung/daVinci/.

Hopcroft, John E. 1971. An n log n algorithm for minimizing the states in a finite automaton. In: Z. Kohavi, editors, *The Theory of Machines and Computations.* Academic Press, pages 189–196.

Hopcroft, John E. and Jeffrey D. Ullman. 1979. *Introduction to Automata Theory, Languages and Computation.* Addison Wesley.

Kaplan, Ronald M. and Martin Kay. 1994. Regular models of phonological rule systems. *Computational Linguistics,* 20(3):331–378.

Lang, Bernard. 1989. A generative view of ill-formed input processing. In *ATR Symposium on Basic Research for Telephone Interpretation (ASTI),* Kyoto Japan.

van Noord, Gertjan. 1995. The intersection of finite state automata and definite clause grammars. In *33th Annual Meeting of the Association for Computational Linguistics,* MIT Boston. Available from http://www.let.rug.nl/~vannoord/papers/.

Oerder, Martin and Hermann Ney. 1993. Word graphs: An efficient interface between continuous-speech recognition and language understanding. In *ICASSP Volume 2,* pages 119–122.

Ousterhout, John K. 1994. *Tcl and the Tk Toolkit.* Addison Wesley.

Pereira, Fernando C. N. and Michael D. Riley. 1996. Speech recognition by composition of weighted finite automata. Available as cmp-lg/9603001 from http://xxx.lanl.gov/cmp-lg.

Roche, Emmanuel and Yves Schabes. 1995. Deterministic part-of-speech tagging with finite-state transducers. *Computational Linguistics,* 21(2).

Sander, G. 1995. Graph layout through the VCG tool. In R. Tamassia and I.G. Tollis, editors, *Graph Drawing, DIMACS International Workshop GD '94, Proceedings; Lecture Notes in Computer Science 894.* Springer Verlag, pages 194–205. cf. also http://www.cs.uni-sb.de/RW/users/sander/html/gsvcg1.html.

Voutilainen, Atro and Pasi Tapanainen. 1993. Ambiguity resolution in a reductionist parser. In *Sixth Conference of the European Chapter of the Association for Computational Linguistics,* Utrecht.

Watson, Bruce W. 1995. *Taxonomies and Toolkits of Regular Language Algorithms.* Proefschrift Eindhoven University of Technology.

A New Quadratic Algorithm to Convert a Regular Expression into an Automaton*

J.-L. Ponty, D. Ziadi, and J.-M. Champarnaud

Laboratoire d'Informatique de Rouen
Faculté des Sciences et des Techniques
76821 Mont-Saint-Aignan Cedex, FRANCE
E-mail:{Jean-Luc.Ponty, Djelloul.Ziadi, Jean-Marc.Champarnaud}@dir.univ-rouen.fr

Abstract. We present a new sequential algorithm to convert a regular expression into its Glushkov automaton. This conversion runs in quadratic time, so it has the same time complexity as the Brüggemann-Klein algorithm and the Chang and Paige one. It provides, however, a representation of the Glushkov automaton that needs only linear space.

1 Introduction

As part of the development of AUTOMATE package [CH91] we try to obtain efficient algorithms, notably to convert a regular expression into a finite automaton. Watson [Wats93] taxonomy is an excellent reference for such a topic. In this paper we are interested in algorithms leading to the Glushkov automaton, which is a natural representation of a regular expression [BS86].

Given E a regular expression we shall denote by n the number of symbol occurrences in E. A straightforward construction of the Glushkov automaton of a regular expression leads to a $O(n^3)$ time complexity [Glus61]. Brüggemann-Klein [B-K93] has shown that this complexity can be reduced to $\Omega(n^2)$. Her algorithm is based on the transformation of E in an equivalent regular expression E^*, called the star normal form (SNF) of E. Chang and Paige [CP92] have used a different data structure, yielding a representation of the Glushkov automaton in time $O(n)$, and a quadratic conversion.

First we recall some definitions and introduce some notations. Next we present the basic Glushkov algorithm [Glus61], and mention the improvements due to Brüggemann-Klein and to Chang and Paige. Last, we describe a new algorithm based on the structure of the parsing tree of the expression E. It yields an original representation of the Glushkov automaton, using forests of states. The interest of this data structure is that it is computed in linear time thanks to a recursive algorithm. Thus, the transitions table is deduced in time $\Omega(n^2)$.

---

* An extended version is to appear in the Belgian Mathematical Society Bulletin, under the title : "Passage d'une expression rationnelle à un automate fini non déterministe".

2 Definitions and notations

In this section, we introduce the terminology and the notations used in the paper. With few exceptions, these notations can be found in [B-K93].

Let Σ be a non-empty finite set of symbols, called alphabet. Σ^* represents the set of all the words over Σ. The empty word is denoted by ε. The symbols in Σ are represented by the first lowercase letters such as $a, b, c...$ The union ($\cup$), the concatenation product ($\cdot$) and the Kleene star closure (*) are the classical regular operations over the subsets of Σ^*. In order to specify the position of the symbols in the expression, the symbols are indexed following the order of reading. For example, starting from $E = (a + \varepsilon).b.a$ we obtain the subscripted expression $\overline{E} = (a_1 + \varepsilon).b_2.a_3$. Subscripts are also called *positions*. The set of positions of a regular expression E is denoted by $Pos(E)$. χ is the application which maps each position in $Pos(E)$ to the symbol of Σ which appears at this position in E. We denote by $\sigma = \{\alpha_1, ..., \alpha_n\}$, where $n = |Pos(E)|$, the alphabet of $\overline{E}$. If F is a subexpression of E, we denote $Pos_E(F)$ the subset of positions of E which are positions in F. We use the classical definitions on regular languages, regular expressions and finite states automata [BBCh92] [AU92].

3 The Glushkov automaton of an expression

In order to construct a non-deterministic finite state automaton recognizing $L(E)$, Glushkov stated four functions that we can define over σ as follows:

Definition 1. $First(E) = \{x \in Pos(E) \,|\, \exists u \in \sigma^* : \alpha_x u \in L(\overline{E})\}$

$First(E)$ represents the set of positions that match the first symbol of some word in $L(\overline{E})$.

Definition 2. $Last(E) = \{x \in Pos(E) \,|\, \exists u \in \sigma^* : u\alpha_x \in L(\overline{E})\}$

$Last(E)$ represents the set of positions that match the last symbol of some word in $L(\overline{E})$.

Definition 3. $Follow(E, x) = \{y \in Pos(E) \,|\, \exists v \in \sigma^*, \exists w \in \sigma^* : v\alpha_x\alpha_y w \in L(\overline{E})\}$

$Follow(E, x)$ represents the set of positions that follow the position x in some word in $L(\overline{E})$.

Definition 4. The function $Null_E$ is defined as $\{\varepsilon\}$ if ε belongs to $L(\overline{E})$ and $\emptyset$ otherwise.

These functions allow to compute the *"Glushkov automaton"* of E, directly defined as $M_E = (Q, \Sigma, \delta, s_I, F)$ with,

1. $Q = Pos(E) \cup \{s_I\}$
2. $\forall a \in \Sigma, \delta(s_I, a) = \{x \in First(E) \,|\, \chi(x) = a\}$

3. $\forall x \in Q, \forall a \in \Sigma, \delta(x, a) = \{y \mid y \in Follow(E, x) \text{ and } \chi(y) = a\}$

4. $F_E = Last(E) \cup Null_E \cdot \{s_I\}$

Notice that Berstel and Pin [BP95] present the recursive computing of the functions First, Last and Follow within the context of local languages.

A basic implementation of the Glushkov algorithm is the following :

- Build the parse tree $T(E)$ of the expression E. Each node ν of this tree represents a subexpression E_ν. To each node ν associate the sets $First(\nu)$, $Last(\nu)$, $Follow(\nu, x)$ and $Null_\nu$.
- For each node ν recursively compute these sets, using the following algorithm (where the left child and the right child of a node ν labelled '+' or '·' are respectively named ν_l and ν_r and where the child of a node ν labelled '*' is named ν_c.)

The Glushkov algorithm

```
switch (ν)
    case ∅ :      First(ν) = ∅; Last(ν) = ∅; Null_ν = ∅;
    case ε :      First(ν) = ∅; Last(ν) = ∅; Null_ν = {ε};
    case x :      First(ν) = {x}; Last(ν) = {x}; Null_ν = ∅;
                  Follow(ν, x)= ∅;
    case '+' :    First(ν) = First(ν_r) ∪ First(ν_r);
                  Last(ν)  = Last(ν_r) ∪ Last(ν_r);
                  Null_ν   = Null_ν_r ∪ Null_ν_r;
    case '·' :    First(ν) = First(ν_r) ∪ Null_ν_r.First(ν_r);
                  Last(ν)  = Last(ν_r) ∪ Null_ν_r.Last(ν_r);
                  Null_ν   = Null_ν_r ∩ Null_ν_r;
                  for each x in Last(ν_r) do
                         Follow(ν, x) = Follow(ν_r, x) ∪ First(ν_r);
                  enddo
    case '*' :    First(ν) = First(ν_c);
                  Last(ν)  = Last(ν_c);
                  Null_ν   = {ε};
                  for each x in Last(ν_c) do
                         Follow(ν, x) = Follow(ν_c, x) ∪ First(ν_c);      [*]
                  enddo
endswitch
```

4 Different implementations of Glushkov algorithm

4.1 A basic cubic algorithm

In Glushkov algorithm, all the unions, excepted for the union labelled [*], are disjoint and can be computed in constant time, using sets represented by lists. If lists are ordered the union labelled [*] can be obtained in linear time, which yields a basic implementation in $O(n^3)$ time.

4.2 Two quadratic algorithms

The union labelled [*] can be computed in constant time if $Follow(\nu, x)$ sets are computed by one of the two following formulas :

- $Follow(\nu, x) = [Follow(\nu_c, x) \setminus First(\nu_c)] \bigcup First(\nu_c)$ (1)
- $Follow(\nu, x) = Follow(\nu_c, x) \bigcup [First(\nu_c) \setminus Follow(\nu_c, x)]$ (2)

These two approaches lead to the Brüggemann-Klein [B-K93] algorithm (formula 1) and to the Chang and Paige [CP92] algorithm (formula 2), which are both in $\Omega(n^2)$ time.

5 A new quadratic algorithm

We present in this section an original algorithm which computes a representation of the Glushkov automaton M_E of a regular expression E in linear time and space. In this algorithm, the computation of the First sets and Last sets is performed on two forests based on the parse tree $T(E)$. The transition function δ of the automaton M_E is represented by a set of links going from the *forest of Lasts* to the *forest of Firsts*. In fact, a link is a cartesian product $L \times F$ where L is a Last set and F is a First set.

5.1 Computation of First sets

Let $T(E)$ be the parse tree of the regular expression E. To each node ν we associate the subtree of root ν and the set of leaves of this subtree, called $Pos(\nu)$. This set can be represented by a linked list. Each node ν, even if a leaf, has a pointer to its leftmost leaf and to its rightmost leaf, so that $Pos(\nu)$ is accessible in constant time.

From the tree $T(E)$, we build the forest $TF(E)$, which represents the First sets defined by Glushkov, as follows :

- $TF(E)$ is initialized by a copy of $T(E)$; each node ν of $T(E)$ is renamed φ in $TF(E)$
- for each node φ labelled ".", we remove the link to its right child φ_r if the empty word does not belong to $L(E_{\nu_l})$ (fig. 1), in order to satisfy the statement : $First(F.G) = First(F) \cup Null_F.First(G)$
- for each node φ :
  - we update the leftmost pointer and the rightmost pointer
  - we link the rightmost leaf of its left child to the leftmost leaf of its right child

After removing of links and updating of pointers, we have the forest $TF(E)$. For each node φ of this forest, the set $Pos(\varphi)$ is exactly the set First of the subexpression E_ν. In a given tree of $TF(E)$, leaves are linked so that access to $First(E_\nu)$ is achieved in constant time.

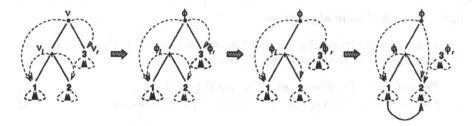

Fig. 1. Updating of the links while computing Firsts

5.2 Computation of Last sets

We build the forest $TL(E)$ of Last sets in the same way as we build the forest of First sets. Each node ν of $T(E)$ is renamed λ in $TL(E)$. In $TL(E)$, for each node λ labelled ".", we remove the link to its left child λ_l if the empty word does not belong to the language $L(E_{\nu_r})$, in order to satisfy the statement : $Last(F.G) = Last(G) \cup Null_G.Last(F)$. As for the forest $TF(E)$, for each node λ we update the different pointers. To each node λ of the forest $TL(E)$ we associate a set $Pos(\lambda)$ which is exactly the Last set of the subexpression E_ν. In a tree of the forest $TL(E)$, leaves are linked so that access to $Last(E_\nu)$ is achieved in constant time.

5.3 Computation of $First(E)$ and $Last(E)$

Let us remark that if ν_0 is the root of the tree $T(E)$, then $First(E)$ is the set $Pos(\varphi_0)$ in the forest $TF(E)$ and $Last(E)$ is the set $Pos(\lambda_0)$ in the forest $TL(E)$.

Lemma 5. *For every regular expression E of size n, we can compute $First(E)$ (resp. $Last(E)$) in time $O(n)$ taking up $O(n)$ space.*

5.4 Computation of δ

Representation of δ. δ is represented by a set of links from the forest $TL(E)$ to the forest $TF(E)$. δ is the set of edges of the Glushkov automaton of the expression E. We compute δ inductively on the tree $T(E)$ from the two following sets :

- Δ_ν is the set of edges computed while evaluating the node ν.
- D_ν is the set of edges computed before evaluating the node ν.

We have the following formulas :

$$\Delta_\nu = \begin{cases} Last(E_{\nu_l}) \times First(E_{\nu_r}) & \text{if } \nu \text{ is labelled } \cdot \\ Last(E_{\nu_c}) \times First(E_{\nu_c}) & \text{if } \nu \text{ is labelled } * \\ \emptyset & \text{otherwise} \end{cases}$$

$$D_\nu = \begin{cases} \emptyset & \text{if } \nu \text{ is a leaf} \\ D_{\nu_c} \cup \Delta_\nu & \text{if } \nu \text{ is labelled } * \\ D_{\nu_l} \cup D_{\nu_r} \cup \Delta_\nu & \text{otherwise} \end{cases}$$

$$\delta = D_{\nu_0} \cup (\{s_I\} \times First(E))$$

Proposition 6. $\delta = \left(\bigcup_{\nu \in T(E)} \Delta_\nu \right) \cup \left(\{s_I\} \times First(E) \right)$

Proof is by induction on the size of E.

Each not empty Δ_ν is a cartesian product $Last(E_{\nu'}) \times First(E_{\nu''})$ represented by a link (λ', φ'') from the node λ' of the forest $TL(E)$ (corresponding to ν') to the node φ'' of the forest $TF(E)$ (corresponding to ν''). So δ is represented by the set L of all these links. Adding a link to L takes a constant time; the links are obtained via a recursive traversal of $T(E)$. So L is computed in linear time.

Removing the redundant links. Let $\Delta_\nu \neq \emptyset$. If there is $\Delta_{\nu'}$ such that $\Delta_\nu \subseteq \Delta_{\nu'}$, the link representing Δ_ν is called redundant. In the example of the figure 2 we have : $\Delta_{\nu_1} \subset \Delta_{\nu_0}$, $\Delta_{\nu_2} \subset \Delta_{\nu_0}$ and $\Delta_{\nu_4} \subset \Delta_{\nu_0}$. So Δ_{ν_1}, Δ_{ν_2} and Δ_{ν_4} are redundant. Removing the redundant links allows to compute δ by means

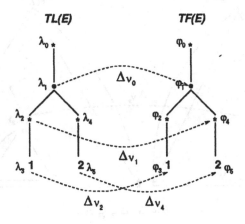

Fig. 2. Computation of Δ_ν for the expression $E = (a_1^* b_2^*)^*$

of a disjoint union. The two following propositions are used to characterize the redundant links. We shall say that a node s' is a descendant of the node s if and only if they belong to the same forest and s' belongs to the tree rooted on s.

Proposition 7. *Let ν and ν' be two nodes in $T(E)$, and φ and φ' the corresponding nodes in $TF(E)$. Then :*

$$\begin{cases} \text{if } \varphi' \text{ is a descendant of } \varphi : & First(E_{\nu'}) \subseteq First(E_\nu) \\ \text{otherwise :} & First(E_{\nu'}) \cap First(E_\nu) = \emptyset \end{cases}$$

Let ν be a node of $T(E)$, λ (resp. φ) the corresponding node in $TL(E)$ (resp. $TF(E)$). Let us assume $\Delta_\nu \neq \emptyset$. Let us denote by λ' the node of $TL(E)$ corresponding to the child of ν if ν is labelled '*', or to the left child of ν if ν is labelled '·'. Let us denote by $F(\lambda')$ the node of $TF(E)$ corresponding to the child of ν if ν is labelled '*', or to the right child of ν if ν is labelled '·'. So, Δ_ν is the set $Pos(\lambda') \times Pos(F(\lambda'))$. With the data structure used, Δ_ν is represented by a link from λ' in $TL(E)$ to $F(\lambda')$ in $TF(E)$.

Proposition 8. *Let ν_1 and ν_2 be two nodes of $T(E)$. Let λ_1 and λ_2 be the two corresponding nodes in $TL(E)$. Let λ'_1 and λ'_2 defined as previously. Let $F(\lambda'_1)$ and $F(\lambda'_2)$ defined as previously.*

$$\text{We have}: \Delta_{\nu_2} \subseteq \Delta_{\nu_1} \Leftrightarrow \begin{cases} \lambda'_2 \text{ is a descendant of } \lambda'_1 \text{ in } TL(E) \\ F(\lambda'_2) \text{ is a descendant of } F(\lambda'_1) \text{ in } TF(E) \end{cases}$$

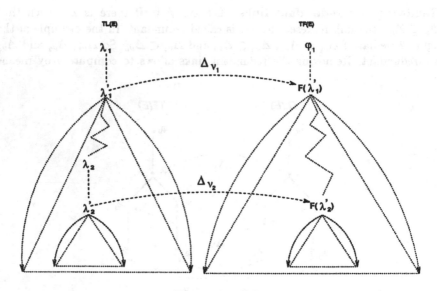

Fig. 3. $\Delta_{\nu_2} \subseteq \Delta_{\nu_1}$

The algorithm <u>Removing</u> which deletes the redundant links is straightforward deduced from the proposition 8. Let us explain some useful points in order to understand the algorithm :

The links $(\lambda'_1, F(\lambda'_1))$ and $(\lambda'_2, F(\lambda'_2))$ appearing in proposition 8 are the links $(a, F(a))$ and $(b, F(b))$ of the algorithm; a and b are nodes of $TL(E)$ or are not defined (nil).

The function $D(x, y)$ tests if x is a descendant of y in the forest $TF(E)$; x and y are nodes of $TF(E)$ or are not defined. We have : $D(\text{nil}, \text{nil}) = D(\text{nil}, y) = D(x, \text{nil}) = $ false.

The functions $l(a), r(a), c(a)$ yield the left child, the right child or the child (closure case) of the node a of $TL(E)$.

The algorithm allows to remove the redundant links in a given tree of the forest $TL(E)$. The exploration of the forest is such that b is always a descendant of a.

The algorithm will be applied on each tree of the forest $TL(E)$; initially b will be the root of the tree, and a will be nil.

The removing algorithm

```
Removing(b,a)
begin
  if b ≠ nil then
    if D(F(b),F(a)) then
      F(b) ← nil            (delete the link (b,F(b)))  ⎫
      switch (b)                                        ⎬ Δ_b ⊆ Δ_a
      case '·','+',x : Removing(l(b),a)                 ⎪
                       Removing(r(b),a)                 ⎪
      case '*' :       Removing(c(b),a)                 ⎭
    else if F(b) ≠ nil then                                  ⎫
         switch (b)                                      ⎫   ⎪
         case '·','+',x : Removing(l(b),b)              ⎬Δ_b≠∅⎪
                          Removing(r(b),b)              ⎪   ⎪
         case '*' :       Removing(c(b),b)              ⎭   ⎬ Δ_a ∩ Δ_b = ∅
    else switch (b)                                     ⎫   ⎪
         case '·','+',x : Removing(l(b),a)              ⎬Δ_b=∅⎪
                          Removing(r(b),a)              ⎪   ⎪
         case '*' :       Removing(c(b),a)              ⎭   ⎭
    endif
   endif
  endif
end
```

Lemma 9. *The representation of δ by a set of non-redundant links can be computed in time $O(n)$.*

From the forests to the Glushkov automaton. We have to build the transition table of the automaton M_E. Edges going from the initial state s_I are deduced from the set $First(\nu_0)$. During the traversal of the tree $T(E)$, every time we meet a link $Last(\nu) \times First(\nu')$, we set $t[i,j]$ to 1, $\forall i \in Last(\nu), \forall j \in First(\nu')$. Since the links are non-redundant, this operation happens exactly once for each edge. So this step takes a time linear on the number of edges, that is to say $\Omega(n^2)$.

Lemma 10. *Computing the transition table from the representation of δ by a set of non-redundant links, is in time $\Omega(n^2)$.*

5.5 Algorithm

We present now the most important steps of the algorithm for computing the Glushkov automaton of a regular expression.

1. Building of the tree $T(E)$ and initialization of its nodes $\qquad$ $O(n)$
2. Building of the forests $TL(E)$ and $TF(E)$ $\qquad$ $O(n)$
3. Computation of the sets Δ_ν $\qquad$ $O(n)$
4. Removing of the redundant links $\qquad$ $O(n)$
5. Transformation into the transition table $\qquad$ $\Omega(n^2)$

Theorem 11. *We can compute the Glushkov automaton M_E of the regular expression E in time $\Omega(n^2)$.*

5.6 Example of conversion

The following example (figure 4) illustrates the main steps for building the Glushkov automaton of a regular expression. The set of edges of the automaton (figure 5) is yielded by the union of disjoint sets Δ_{ν_i}, augmented by edges going from the initial state to states of the set $First(\nu_0)$.

$$\delta = \Delta_{\nu_2} \cup \Delta_{\nu_1} \cup \Delta_{\nu_0} \cup (\{s_I\} \times First(\nu_0))$$

$$\delta = \{(0,1), (0,2), (0,3), (1,1), (1,2), (2,1), (2,2), (1,3), (2,3), (3,4)\}$$

Final states of the automaton are states of the set $Last(\nu_0)$, augmented by the initial state if the empty word belongs to $L(E)$.

6 Conclusion

Given E a regular expression, and n the number of symbol occurrences in E, our algorithm computes in time $O(n)$ an original representation of M_E, the Glushkov automaton of E. This representation is based on two forests and is such that the transition function δ is implemented by a collection of links going from one forest to the other. Every link represents the cartesian product of two sets of positions of E. A given edge of M_E may appear inside several links. We have presented an $O(n)$ time procedure computing a subset of non redundant links (that is every edge of M_E appears in one and only one of these links). This procedure is close to the Star Normal Form (SNF) defined by Brüggemann-Klein since the set of links directly constructed for a SNF expression is non redundant. Let us add that our algorithm has been parallelized on a PRAM model [DZ-JMC95] [DZ96].

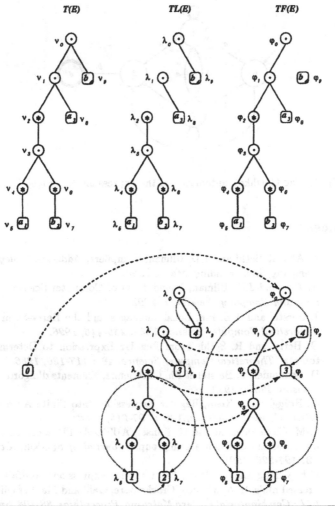

Fig. 4. Example of conversion with the expression $E = (a_1^* b_2^*)^* . a_3 . b_4$

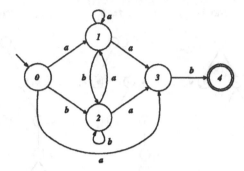

Fig. 5. The Glushkov automaton of the expression $E = (a_1^* b_2^*)^*.a_3.b_4$

References

[ASU86]   A. Aho, R. Sethi and J-D. Ullman, Compilers, *Addison-Wesley Publishing Company, Inc., Reading, Mass., 1986.*

[AU92]   A. Aho and J.D. Ullman, Foundations of Computer Science, *W.H. Freeman and Company, New York, 1992.*

[BP95]   J. Berstel and J-E. Pin, Local languages and the Berry-Sethi algorithm, *Theoretical Computer Science,* **155** : *439-446, 1996.*

[BS86]   G. Berry and R. Sethi, From Regular Expression to Deterministic Automata, *Theoretical Computer Science,* **48** : *117-126, 1986.*

[BBCh92]   D. Beauquier, J. Berstel et Ph. Chrétienne, Eléments d'algorithmique, *Ed. Masson, Paris, 1992.*

[B-K93]   A. Brüggemann-Klein, Regular Expressions into Finite Automata, *Theoretical Computer Science,* **120** : *197-213, 1993.*

[CH91]   J.-M. Champarnaud and G. Hansel, AUTOMATE, a computing package for automata and finite semigroups, *Journal of Symbolic Computation,* **12**, *197-220, 1991.*

[CP92]   C.H. Chang and R. Paige, From regular expressions to dfa's using compressed nfa's. In Apostolico, Crochemore, Galil and Manber editors, *LNCS 644 : Combinatorial Pattern Matching, Proceedings, 88-108, Springer Verlag, 1992.*

[Glus61]   V.-M. Glushkov, The abstract theory of automata, *Russian Mathematical Surveys,* **16**, *1-53, 1961.*

[Wats93]   B.W. Watson, Taxonomies and Toolkits of Regular Language Algorithms, *CIP-DATA Koninklijke Bibliotheek, Den Haag, Ph. D., Eindhoven University of Technology, 1995.*

[DZ96]   D. Ziadi, Algorithmique parallèle et séquentielle des automates, Thèse de doctorat, Université de Rouen, 1996.

[DZ-JMC95] D. Ziadi and J.-M. Champarnaud, An optimal parallel algorithm to convert a regular expression into its Glushkov automaton, *accepted in Theoretical Computer Science, rapport LIR95.10 Informatique Fondamentale,* Université de Rouen, 1995.

Implementing Sequential and Parallel Programs for the Homing Sequence Problem

B. Ravikumar and X. Xiong

Department of Computer Science
University of Rhode Island
Kingston, RI 02881
{ravi,xiong}@cs.uri.edu

Abstract. Homing sequences play an important role in the testing of finite state systems and have been used in a number of applications such as hardware fault-detection [7], protocol verification [4], and learning algorithms [11, 3, 1] etc. Here we present a parallel program implementation that finds a homing sequence for an input DFA. Our program can handle *randomly* generated instances with millions of states, and *all* DFA's with thousand states. In addition to the design, analysis and implementation of the algorithm, we also discuss what constitute good test cases to test programs that deal with finite automata.

1 Introduction

In most applications involving finite automata, the underlying DFA is a black-box whose internal states are not accessible. Testing problems involve determining if the DFA has a certain property using I/O experiments. An I/O experiment consists of applying an input and observing the output. Before performing a test, the DFA should be brought to a known state. An input sequence x with the property that the output on x uniquely determines the state reached (after applying x) is known as a *(preset) homing sequence*. For example, 010 is a homing sequence for the machine described in Figure 1. Table 1 shows the machine's response to the sequence. Note that although 010 produces the same outputs from states C and D, the definition of a homing sequence is not violated since the same state is reached in both cases. However, 01 is not a homing sequence since the outputs produced from states A and B on 01 are the same, but they reach two distinct states after reading this input. Finding a homing sequence in a finite automaton is an important problem in the testing of finite state machines and has been studied and used extensively since the mid 1950's. Some recent applications of homing sequences can be found in the following list of papers: [11, 7, 3, 1]. These applications span a wide range of topics: exploration of an unknown environment [11, 3, 1], hardware fault-detection [7], protocol verification [4] etc.

Every minimal finite automaton with n states has a homing sequence of length $O(n^2)$. It is possible to find such a sequence in time $O(n^3)$. In recent applications, the DFA's for which a homing sequence is required have several thousands of states. Such instances are hard to handle on a uniprocessor machine both because of the space as well as time requirements.

Table 1. Response to the Sequence 010

Initial State	Response	Final State
A	000	A
B	001	D
C	101	D
D	101	D

A carefully designed parallel program is not only desirable but perhaps the only way to handle such instances. Our goal is to design a parallel algorithm with provably good time complexity which can also be implemented on a real parallel machine. (Even some of the sequential programs used in practice have not been analyzed, or are not guaranteed to work on all inputs.)

With this goal, we designed two parallel algorithms for this problem. The two algorithms use very different techniques. One of them is a deterministic algorithm of time complexity $O(\sqrt{n}log^2 n)$ and the other one is a randomized parallel algorithm [10]. The randomized algorithm is an RNC algorithm. Both algorithms are theoretically efficient, but they are not useful in practice for several reasons: the RNC algorithm has a large processor requirement (about $O(n^6)$) which clearly limits the size of the inputs it can handle. The deterministic algorithm is very complicated and is communication intensive. Neither algorithm can be implemented without some drastic simplification and/or modification of goals (such as good performance *on average* rather than in the worst-case). Here we present a simple parallel program implementation for this problem.

First, we will state succinctly the objectives of our implementation. Every

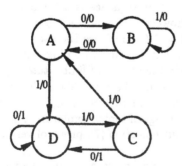

State	Next State,Output input=0	input=1
A	B,0	D,0
B	A,0	B,0
C	D,1	A,0
D	D,1	C,0

Fig. 1. A DFA Example

DFA has many homing sequences and in fact, a very long random sequence is almost surely a homing sequence. (Often, even a short random sequence is a homing sequence, but this is not true when the DFA is selected by an adversary.) Thus our problem is actually an optimization problem; clearly, shorter homing sequences are more desirable. Unfortunately, finding the shortest length homing sequence is NP-hard [11], and NP-hardness holds even when restricted to special cases such as permutation DFA's [8]. Thus any homing sequence of length bounded by polynomial in n, the number of states, is acceptable.

The parallel computers on which our algorithms have been implemented are IBM SP-2 and CM-5. These are *coarse-grained* parallel computers. The number of processors is relatively small (512 or 1024), and each processor is a powerful stand-alone workstation. We require our parallel program to satisfy the following criteria:

(i) The program should work on *very large* size randomly chosen input DFA's. Our goal is to handle DFA's with close to 1 million states. It seems unlikely that *every* DFA with such large size can be handled even by a parallel program running on the fastest and the largest super computer, since the size of the output itself is of the order of 10^{12}. On the other hand, we would like our parallel program to be able to handle *every input DFA with a few thousand states.* These two cases (random instances of the order 10^6 and every instance of order 10^3) are already beyond the limits of serial programs. The main reason is the lack of memory to store all the data structures needed to implement the algorithm. We require our parallel program to push the limit on the size of the input instances that can be handled both in the worst-case as well as in the typical case.

(ii) Our programs should handle *implicit* representation of DFA's. i.e., the transition function δ may not be given as a table, but can be accessed by calling a function. This representation is useful in hardware testing applications in which the DFA is defined by a sequential circuit.

(iii) The program should scale well. This means that the total number of operations performed (including the communication and other message exchange operations) by the program should not drastically increase when the number of processors used increases.

(iv) The program should exhibit a reasonable speed-up. This means that the same instance should run faster when the number of processors is increased, provided that the instance is not too small. Finally, the program should have predictable performance and be transportable across different parallel machine platforms. We also wanted to study this problem as a test case for determining the practical effectiveness of parallel algorithm design techniques. We are not aware of any other parallel program implementation for this problem, although many sequential programs have been designed and implemented.

In this paper, we will describe a simple parallel program which satisfies most of the above requirements. In section 2, we will present some basic definitions and describe the main idea behind our parallel program. In section 3, we describe the implementation details and analyze the time and communication complexity of our program. In section 4, we present the experimental performance of our program. We conclude with directions for future work.

2 Definitions and Preliminaries

A finite automaton (DFA) or a finite state system M is a 5-tuple $M = < Q, I, O, \delta, \lambda >$ where Q is a finite set of states, I is a finite set of input symbols, O is a finite set of output symbols, $\delta : Q \times I \rightarrow Q$ is the transition function and $\lambda : Q \times I \rightarrow O$ is the output function. Note that we use the *Mealy machine* model but all our results hold for *Moore machines* as well. Following [11], we will use the convenient abbreviation $q < x >$ to denote $\lambda(q, x)$ and qx to denote $\delta(q, x)$ throughout. If $R \subset Q$ and $x \in \Sigma^*$, we define $\delta(R, x) = \{\delta(r, x) | r \in R\}$. As above, we abbreviate $\delta(R, x)$ by Rx. In a similar way, we define $\lambda(R, x)$ and abbreviate it $R < x >$. A string x is said to be a distinguishing string for two states $p, q \in Q$ if $p < x > \neq q < x >$.

Let $M = < Q, I, O, \delta, \lambda >$ be a DFA. The following definitions are with respect to M. Let $A \subseteq Q$. A preset homing sequence for the collection A is a string x such that for any $p, q \in A$, if $p < x >= q < x >$ then $px = qx$. An adaptive homing sequence for the collection A is a tree T whose internal nodes labeled by symbols of I, whose edges are labeled by symbols of O and each of whose leaves is labeled by a state in Q, with the property that if M starts in any state in A and the input at the root is applied and the edge labeled by the output is traversed and the input at the next node is applied etc., then M will be in the state indicated by the label associated with the leaf reached.

The study of homing sequences was initiated by Moore in his classical work [6]. Early work by him and others showed the existence of a homing sequence of length $O(n^2)$ for any minimal DFA M with n states. From now on, we will assume that the input DFA is minimized. A sequential algorithm (taken from [11]) to construct a homing sequence is presented in *HSEQ*. In line 4, the algorithm requires a distinguishing string x for the states ph and qh. We assume that distinguishing strings for all pairs have been already computed and stored in a table. These strings can be obtained by a modified DFA minimization algorithm and the existence of such a string is guaranteed for all pairs of states by the minimality of M. In fact, such a modified minimization algorithm can find the shortest distinguishing string for all pairs of states in time $O(n^2)$. It is well-known (see e.g. [5]) that for any pair of states in an n state DFA, a distinguishing string of length at most $n - 1$ exists. It is also not difficult to see that the number of iterations of the *while* loop is bounded by $n - 1$, since the quantity $|Q < h >|$ increases by at least 1 after each iteration and $|Q < h >|$ is bounded by n. Thus the length of the output h is bounded by $(n - 1)^2$. In the next section, we will show how this algorithm can be implemented efficiently in practice.

```
procedure HSEQ;
  begin
1.      h ← ε;
2.      while ∃ p, q ∈ Q s.t. p < h >= q < h > and ph ≠ qh do
3.      begin
4.          Let x be a distinguishing string for states ph and qh;
5.          h ← hx;
6.          Q ← Qh;
7.      end
8.      output(h);
  end
```

A Sequential Program

We will modify the above algorithm to compute a homing sequence for a subset $X \subseteq Q$ of states. (Recall the definition of a homing sequence *for a given set X* as give in Section 2.) It is not difficult to see that algorithm *SIMPLE_HSEQ* achieves this. We will first briefly describe how the algorithm *SIMPLE_HSEQ* can be implemented as a serial program on a uniprocessor. At any stage, the program maintains a prefix h of the homing sequence and the set $X < h >$ of states reached on h starting from some state in X. It actually maintains a partition of states in $X < h >$ so that two states $p, q \in X < h >$ belong to the same partition if and only if there exist states $p, q \in X$ so that $p < h >= q < h >$ but $ph! = qh$. With this information structure, it is easy to implement the *while* condition. Any two states belonging to the same class in the partition can be selected arbitrarily. It is not difficult to see that the number of iterations of the *while* loop is bounded by $|X| - 1$ and hence the total time complexity of this algorithm is at most $O(n^2|X|)$.

```
procedure SIMPLE_HSEQ(X);
{Sequential algorithm to find a homing sequence for a set X ⊆ Q}
  begin
1.      h ← ε;
2.      while ∃ p, q ∈ X s.t. p < h >= q < h > and ph ≠ qh do
3.      begin
4.          Let x be a distinguishing string for states ph and qh;
5.          h ← hx;
6.          X ← Xh;
7.      end;
8.      output(h);
  end.
```

Program for Homing a Subset of Q

A high-level description of the parallel algorithm is as follows. Suppose there are p processors $P_1, ..., P_p$. The state set $X = Q$ is divided into $Q_1, Q_2, ...,$ Q_p, each with roughly the same number of states. All processors P_i find a homing sequence for the states Q_i in parallel, but each processor essentially runs the sequential algorithm described in *SIMPLE_HSEQ*. Then, each processor P_i communicates the string h_i it found to other processors. Based on some conditions, the processors select one of the strings h_i, say h', and output it as part of the final output, also determine $Q < h' >$, the states reached on input h' from all possible initial states. The set, after some processing, becomes the new set X of states. It is again divided into subsets and assigned to various processors and another cycle is repeated. When the current set X becomes small enough, we will simply continue the algorithm on a single processor. The concatenation of the sequence of strings produced after each round is the final output.

3 Implementation Details

In this section, we will present the implementation details of the program described in Section 2. We will first present the primitive operations of the parallel machine model. The program was designed based on *message passing* style. In this model, each processor will be assigned data and the same program will be implemented on each. Conditional statements can use the processor ID, however, so that different processors can execute different segments of the program. The processors share information by communicating with each other. Given below is a detailed high-level description of our program. The program has been implemented in C using the communication library CMMD of CM-5.

Program *Parallel_Homing*
Input: A DFA $M = < Q, I, O, \delta, \lambda >$. The program uses k processors (k can be specified). The state set Q is partitioned into $Q_1, Q_2, ..., Q_k$ of equal size. Q_i is assigned to the processor P_i.
Output: A homing sequence of M.
Comment: At each stage, a set of states will be assigned to each processor. The collection of all such states is called the current_set. Further, each state in the current_set also has a key associated with it, namely the class number in the partition of current_set. In the following, we denote the pair (state label, partition number) as the *extended state label*.
Begin

Initialize current_set to $Q0$, where 0 is some symbol in Σ; Initialize the key associated with state p as $\lambda(p, 0)$. Initialize output to empty string;
while (the size of current_set $> k$) **do in parallel**

1. Processor P_i finds a homing sequence h_i for Q_i using the (sequential) algorithm *SIMPLE_HSEQ*.
2. Processors communicate h_i to a single processor, e.g. P_1 which selects one string h' from h_i's using some criteria. (See the discussion below.)
3. P_1 sends h' to all processors. h' is appended to the output list.
4. Processor P_i computes the set $Q_i h'$ of states reached on input h' and the output strings produced $Q_i < h' >$ from states Q_i. (This is done sequentially by each processor.) current_set $= \bigcup Q_i h'$. At this point, processor P_i holds the states $Q_i h'$ and the partition label for

state ph' is assigned as the partition label of p. (I.e. the old label is retained.)

5. Duplicates are removed from current_set. This is done as follows: First current_set is globally sorted using the extended state label as the key. This brings together identical reached states belonging to the same equivalence class of the old partition. Processor P_i scans $Q_i h'$ and removes duplicates. Then, adjacent processors communicate and remove shared duplicates, if any. Let the resulting set be renamed current_set.

6. The partition $Q_j h'$ is refined so that two states p, q belong to the same equivalence class only if there exist $p', q' \in Q_j$ such that $p' < h' > = q' < h' >$ and $p = p'h' \neq q'h' = q$. This results in a new partition label (which could be the old label) for each state in current_set. (This step is elaborated further below.)

7. Singleton members of the new partition are deleted. current_set is set to the resulting collection.

8. current_set is divided into $Q_1, ..., Q_k$.

end_while

Transfer the current_set to a single processor.

Call the sequential program to find a homing sequence h for current_set; Append h to the output list.

end.

We now analyze the main computational steps and the communication operations of the algorithm for each iteration *in the worst-case*. To get the total estimate, we need to multiply it by the number of iterations. We do not know a tight bound on the number of iterations. The obvious bound of n (the number of states) seems too pessimistic, even in the worst-case. But on randomly generated test cases, this number turns out to be a small constant. In what follows, we will present the computational cost and communication cost of each step separately.

Computational Cost of the Program

Since each processor is assigned $\frac{n}{k}$ states, step (1) takes $O(\frac{n^3}{k})$ operations.

In step (2), we can choose any one h_i using some heuristic. For example, we may pick the shortest one since we want a short overall output string. But the rationale for selecting the longest one is equally compelling. The longer ones may home in many states belonging to the other classes of the partition. For random DFA's, experiments indicate that longer h_i's are better. But if h_i is too long, sorting of outputs becomes expensive. So in our implementation, we choose the longest h_i such that the output of each state can be packed into an integer. We then can use a standard sorting subroutine. The cost of step (2) is $O(k)$ comparison operations.

In step (4), processor P_i computes the sets $Q_i h'$ and $Q_i < h' >$. This is done in a straight-forward manner in $O(\frac{n}{k} \cdot |h'|)$ time.

Step (5) is aimed at grouping identical states together in order to remove duplicates. This is done in $O(\frac{n}{k} \cdot \log n)$ time by parallel sorting. Sorting is carried out within each segment of the partition by using the extended state label as the key.

Step (6) is implemented as follows. A global sorting is performed using the pair $(q < h' >,$ partition label) as the key for each $q \in current\_set$. After sorting, every adjacent key pair is examined and a split is made at places where the output strings differ. The complexity for this step is $O(\frac{n \log n}{k})$.

Step (7) is implemented by scanning the list sorted in Step (6) locally by each processor. This is easily seen to take at most $O(\frac{n}{k})$ steps.

In Step (8), the number of remaining states is computed in $O(\log k)$ time by a prefix summation. The redistribution of $current\_set$ is part of the parallel sorting step (6).

The dominant cost of the above is the cost of step (1). If the number of iterations of the main loop of the parallel program is not large, then we would have a speed-up close to p. Although this is not true in the worst-case, it appears to be true when the input instances are randomly generated. In any case, the goal of our program design is to achieve a good performance *in practice* and so the best test of our design is the experimental performance. This is discussed in the next section.

Communication Cost of the Program

The most desirable feature of our program is that it uses very few communication operations. The communication steps are (2), (3) and the communication cost for two parallel sorting steps. The dominant communication cost is for the sorting step. Many communication efficient programs have been designed and we can use any of them. Our implementation is based on a program due to [12].

4 Experimental Results on CM-5

We ran the program presented above for some DFA's of different sizes on a CM-5 with 32 nodes, each node with 4 vector units. Since our main goal is to demonstrate that our program works well *in practice*, the issue of creating good test data is critical. It appears that very little work has been done in collecting good bench marks and test data comprising of generic DFA's. Specific applications such as hardware fault-detection have some bench marks (e.g. ISCAS'89), but these circuits have a large number of outputs and hence are not particularly difficult instances. The worst-case DFA's with n states (those in which the shortest homing sequence is of length $n(n-1)/2$) are known [5]. A standard way to create test data is to generate instances randomly. The limitations of random models are well-known. (Even for some NP-hard problems, there are simple algorithms that solve almost all instances very efficiently.) But in the absence of any real data, random models provide a reasonable alternative. In the case of automata, two random graph models have been studied, mainly by Russian mathematicians. For want of better terminology, we will call them random model and the uniform random model. In the random model, a random instance with n states over input alphabet I and output alphabet O is produced by assigning $\delta(q, a)$ a state with probability $1/n$ and by assigning $\lambda(q, a)$ an output symbol with probability $1/|O|$. The uniform model is defined as follows: δ is selected arbitrarily (i.e., an adversary can select it) but the output function is chosen randomly. A number of interesting results about homing sequences are known for such models. For example, almost all DFA's with n states have a homing sequence of length $O(log\ n)$ for a random model; almost all DFA's with n states have a *local* homing sequence of length $O(log\ n)$ for each state for a uniform random model. (A local homing sequence for a state p is a string x such that based on the output, we can determine if the state reached is px.)

We propose to use a variety of test cases (including all the ones presented above) to test our program. Here we will present the test results only for two cases: (i) randomly created DFA, and (ii) DFA created using the uniform random model in which the δ function is chosen as in the worst-case DFA's. The random instances were created as follows: the input and output alphabet were both of size 2, and the size of the DFA was selected in the range from a few thousands to half a million.

A DFA with equivalent states may not have a homing sequence. So we assume that DFA's to be homed are minimized and thus have no equivalent states. Actually, our program still works if the input DFA is not minimized, but it will not be efficient in such instances. A random DFA generated as above may have equivalent states, but experiments show that this is unlikely.

Figure 2 shows the running time in seconds for the program running randomly generated DFA's as inputs. The time to run the message passing program was taken as elapsed CPU time of one node (provided by the CM-5 CMMD library), but not including the time taken by the sequential algorithm in the last step. This timing for parallel part is reliable since all nodes are synchronized at the beginning and end of the program (and also in each iteration).

To study the scalability of the program, we chose a DFA with 256K states and found a homing sequence for it by using a CM-5 configured with 32, 64, 128 and 256 nodes. Figure 3 shows the results of running time. The program seems to scale well when the number of nodes is smaller than 256. It may also have good scalability for more nodes if the input size becomes large.

We also tested our program with DFA's generated using the uniform random model. Here, the state transition of the DFA is fixed arbitrarily and only the output function is chosen randomly. The most difficult DFA (i.e., the one for which the program takes maximum time) under this model seems to be the one

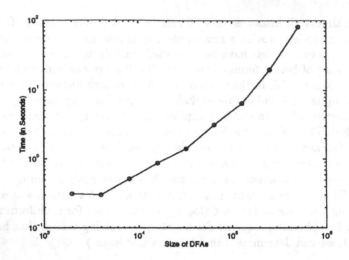

Fig. 2. Running Time vs. Problem Size of the Parallel Program

obtained by fixing the transition function as in the worst-case DFA. When we do this, our program's elapsed CPU time grows significantly larger than that for the random DFA's tested above. Currently our program can handle these DFA's with up to a few thousand states. Going beyond 10^5 in this model may not be easy since in these cases, the output length of even the shortest homing sequence can be 10^{10}.

5 Conclusions

In this paper, we presented a practical parallel program implementation for the homing sequence problem. Until recently, implementing parallel programs with good performance was feasible only for well-structured problems such as vector or matrix computations. Data parallel style offered a simple and effective solution for such problems. Recent advances in the design of parallel programming languages and other system tools, combined with coarse-grained parallel machines, are extending the scope of parallel programs to nonnumerical problems. Our goal is to design a library of parallel programs for all the basic problems related to finite automata. So far we have considered two problems in some detail: DFA minimization [9], and the homing sequence problem. Analyzing these two problems has offered some understanding about the nature of parallel computation for such applications. There is no doubt that parallel programming will become standard in the near future. MPI running on a network of workstations offers a good speed-up at modest hardware cost. Thus it is of interest to develop a parallel program library for all fundamental problems. Our preliminary work

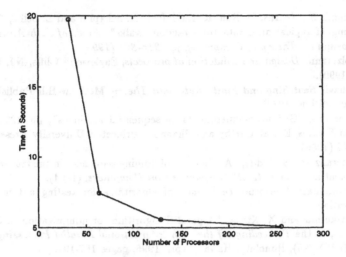

Fig. 3. Scalability of the Parallel Program

along these lines appears to be promising.

There are a number of directions in which this work can be extended. We will list some of them below.

(i) Designing a sequential or a parallel program for the adaptive homing sequence. These are decision trees in which the input symbol to be applied next can depend on the outputs to the previous input symbols. In previous work (e.g. [11]) it has been noted that the adaptive sequences are relatively shorter (although not in the worst-case).

(ii) Designing sequential or parallel programs for synchronizing, distinguishing or testing sequences. These are closely related to homing sequences and are also of interest in automata testing and inference problems.

(iii) Examining the issues of programming language and environment. Recently [2] has argued for the effectiveness of functional language NESL in parallel program design. We would like to look at NESL or similar languages for implementing automata. Also of interest is developing a parallel version of Grail or similar programs.

References

1. M. A. Bender and D. K. Slonim, "The power of team exploration: Two robots can learn unlabeled directed graphs", *35th Annual IEEE Symposium on Foundations of Computer Science*, pp. 75-85 (1994).
2. G. Blelloach and J. Harwick, "Class Notes: Programming Parallel Algorithms", *Technical Report CMU-CS-93-115, Carnegie Mellon University*.

3. Y. Freund, M. Kearns, D. Ron, R. Rubinfeld, R. Schapire and L. Sellie, "Efficient Learning of typical automata from random walks", *Proc. of 25th Annual ACM Symposium on Theory of Computing, pp. 315-324 (1993)*.

4. G. Holzmann, *Design and validation of protocols*, Englewood Cliffs, NJ, Prentice-Hall (1990).

5. Z. Kohavi, *Switching and Finite Automata Theory*, McGraw-Hill Publishers Inc. Second Edition (1978).

6. E. F. Moore, "Gedanken-experiments on sequential machines", pp. 129-153, *Automata Studies*, Ed: McCarthy and Shannon, Princeton University Press, Princeton, NJ (1956).

7. I. Pomeranz and S. Reddy, "Application of homing sequences in the fault-detection of sequential circuits, *IEEE Transactions on Computers*, (1994).

8. B. Ravikumar, "Performance bounds of algorithms for testing automata", (in preparation).

9. B. Ravikumar and X. Xiong, "A parallel algorithm for minimization of finite automata", in the *Proceedings of the 10th International Parallel Processing Symposium* (IPPS '96), Honolulu, Hawaii, April 1996, pages 187-191.

10. B. Ravikumar and X. Xiong, "Randomized parallel algorithms for the homing sequence problem", to appear in *the International Conference on Parallel Processing* (ICPP'96), Bloomingdale, Illinois, August 1996.

11. R. L. Rivest and R. E. Schapire, "Inference of finite automata using homing sequences", *Information and Computation*, vol. 103, pp. 299-347, (1993).

12. A. Tridgell and R.P. Brent, "An Implementation of a General-Purpose Parallel Sorting Algorithm", *Report TR-CS-93-01, Computer Science Laboratory, Australian National University*, February, 1993.

Integrating Hands-on Work into the Formal Languages Course via Tools and Programming

Susan H. Rodger*

Duke University, Box 90129, Durham, NC 27708-0129, USA

Abstract. Integrating hands-on practice into an automata and formal languages course aids in transforming the course from a traditional mathematics course into a traditional computer science course, while making the material more interesting from both teaching and learning perspectives. The interactive and visual tools we integrate into our course are FLAP, a tool for constructing and simulating several types of nondeterministic automata, and LLparse and LRparse, tools for constructing parse tables and animating the parsing of strings. As a programming component, our students are also required to write an LR(1) parser for a simple programming language, using the tool Xtango to animate programs in this new language.

1 Introduction

Successful learning requires some form of feedback, but the usefulness of such feedback can be greatly reduced if delayed. For example, when written homework is not returned immediately, the problems may be forgotten, or even worse, a student may mistakenly continue to build on the misunderstandings. Immediate feedback allows the student to recognize misunderstandings quickly (and thus to seek help from the instructor), enabling confident progression in one's studies.

Traditional approaches to problem solving in automata theory involve tedious solutions written on paper, susceptible to errors, and lengthy delays in feedback. In a textual approach, students write solutions for a finite automaton or a Turing machine by writing formal notation for all parts of the machine, including a table of transitions. In a graphical approach, students write solutions in the form of a transition diagram. Our experience has been that both approaches are error prone, in the same manner that a program is usually not correct the first time it is compiled and run. Furthermore, answers on paper generally take a long time to grade and return.

Previous studies suggest that hands-on practice with animations in algorithm courses aids in understanding. Although such animations clearly excite students, the animations only slightly assisted student understanding [2]. However, allowing students to build their own animation led to a higher understanding of an algorithm [3]. We take a similar approach with the automata theory course.

---

* Supported in part by the National Science Foundation's Division of Undergraduate Education through grants DUE-9596002 and DUE-9555084.

This paper explains how we have integrated hands-on practice with immediate feedback into CPS 140, the undergraduate automata and formal languages course at Duke University. Immediate feedback is achieved through visual and interactive tools and programming assignments. Tools are used during lectures by the instructor [9] for explaining concepts, solving problems, and as an introduction to using the tool; and in labs and homework by the student to solve problems. The tools we use include FLAP [8, 5], a tool for constructing and simulating several types of nondeterministic automata, and LLparse and LRparse [4], instructional tools for constructing parse tables and animating the parsing of strings. For the programming component, our students write an LR(1) parser for a simple programming language and use the tool Xtango [10] to animate programs in this language.

In Section 2, we describe the tool FLAP and describe how we have integrated it into the course CPS 140. In Section 3, we describe the tools LLparse and LRparse and their use in CPS 140. In Section 4, we describe the LR(1) parser programming assignment and the use of Xtango for animating programs. Section 5 describes the evaluation of these tools, and Section 6 gives concluding remarks.

2 FLAP

In this section we describe FLAP [8, 5], and several assignments and lectures using FLAP.

2.1 Overview of FLAP

FLAP is a visual tool for designing and simulating several variations of finite automata (FA), pushdown automata (PDA), Turing machines (TM), and two-tape Turing machines (TTM), including nondeterministic versions of these machines. In the building window of FLAP, one draws a graphical representation of the transition diagram. States and arcs between states can be easily created using the mouse; in the latter case, a transition label appears that can be filled in by the user. There are options for making a state final or nonfinal, moving a state or label, deleting a state or label, saving or retrieving machines, and identifying nondeterministic states.

Once the drawing of the automaton is complete, one can simulate it by entering an input string and selecting either fast or slow (step-by-step) simulation. In the fast simulation, a complete tree of configurations is generated, with the root containing the start state, input string and any other relevant information. Although this configuration tree is not displayed, a message informs the user whether or not the input string was accepted (i.e. if some path in the tree leads to acceptance). If the string is accepted, the user can choose to step through an animation of the configurations on the path that led to acceptance.

For example, the animation for a PDA initially shows a picture of the starting configuration containing the start state, the input string and an empty stack.

Each step shows the new state, the symbols remaining to be processed in the input string, and the current contents of the stack.

In the slow simulation, the user steps through each level of configurations generated in the tree at his or her own pace, starting with the start configuration. No more than 12 configurations are displayed at any time. If more than 12 configurations exist at any level, the user must select configurations to remove or freeze. In the latter case, they cannot be expanded until they are thawed. At any point in a simulation, a particular configuration can be selected and traced to see how it was reached.

FLAP has been designed with a common interface so that once a student has studied one type of automaton, it is easy to construct another type. Since the definitions of these machines are general, several variations of each machine can be examined, allowing the tool to be used along with most automata theory textbooks [6, 7].

2.2 Definitions of Automata

FLAP recognizes a general definition of each type of automaton so that several variations of each can be studied.

Definition 1. A nondeterministic finite automaton M is represented by the 5-tuple $M = (Q, \Sigma, \delta, q_0, F)$, where Q is a finite set of states, Σ is a finite set of input symbols, δ is a set of transitions represented by $\delta : Q \times \Sigma^* \to 2^Q$, q_0 is the start state ($q_0 \in Q$), and F is a set of final states ($F \subseteq Q$).

Since this definition is general, other variations of FA can be examined, including a deterministic FA (DFA). Another variation restricts the number of input symbols that can be processed in a transition to one symbol. In all variations, a FA accepts an input string if there is a path from the start state to a final state that recognizes this input.

Definition 2. A nondeterministic pushdown automaton M is represented by the 7-tuple $M = (Q, \Sigma, \Gamma, \delta, q_0, Z, F)$, where Q is a finite set of states, Σ is a finite set of tape symbols, Γ is a finite set of stack symbols, δ is a set of transitions represented by $\delta : Q \times \Sigma^* \times \Gamma^* \to$ finite subsets of $Q \times \Gamma^*$, q_0 is the start state ($q_0 \in Q$), Z is the start stack symbol ($Z \in \Gamma$), and $F \subseteq Q$ is a set of final states.

This definition is general so that variations of a PDA can be examined, including deterministic PDA, restricting the number of input symbols to one or zero (λ) symbols to be processed, and restricting the number of symbols popped on each transition to exactly one. There are two definitions for the acceptance of an input string, providing additional variations of PDA that can be studied. Acceptance is based on either reaching a final state or the stack becoming empty.

Definition 3. A nondeterministic Turing machine M is represented by the 7-tuple $M = (Q, \Sigma, \Gamma, \delta, B, q_0, F)$, where Q is a finite set of states, Σ is the input alphabet, Γ is the tape alphabet (with $\Sigma \subseteq \Gamma - \{B\}$), δ is a set of transitions represented by $\delta : Q \times \Gamma \to 2^{Q \times \Gamma \times \{R,L,S\}}$, B is a special symbol denoting a blank on the tape, q_0 is the start state ($q_0 \in Q$), and $F \subseteq Q$ is a set of final states. The symbols R, L and S denote directions Right, Left, and Stay.

Other variations of the TM are a deterministic TM, a two-tape TM (δ is represented by $\delta : Q \times \Gamma \times \Gamma \to 2^{Q \times \Gamma \times \Gamma \times \{R,L,S\} \times \{R,L,S\}}$) and a variation when the movement of the tape head is restricted to always moving, either R or L. In all cases, acceptance of an input string is based on reaching a final state.

In FLAP, the transition format for all machines is:

> FA: $< in\_string >$
> PDA: $< in\_string >$, $< pop\_string >$; $< push\_string >$
> TM: $< in\_symbol >$; $< write\_symbol >$, $< dir\_symbol >$

where $< in\_string >$ is the (zero, one or more) symbols that must be present on the input tape if the transition is to be applied, $< pop\_string >$ is the symbols that must be present at the top of the stack which are to be popped off, $< push\_string >$ is the symbols that are to be pushed on the stack, $< in\_symbol >$ is a single symbol which must be the current symbol on the input tape if the transition is to be taken; $< write\_symbol >$ is the single symbol which is written to the current position on the input tape; and $< dir\_symbol > \in \{R, L, S\}$, depending on whether the read head should move Right, move Left, or Stay put. For $< pop\_string >$ and $< push\_string >$, the leftmost symbol corresponds to the top of the stack. The two-tape TM has two connected labels (one for each tape), both in the TM format above.

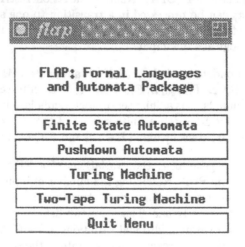

Fig. 1. FLAP menu

2.3 Examples using FLAP

An NPDA Example. In this section we first show an example of constructing and simulating a nondeterministic PDA, and then give brief examples of other automata constructed using FLAP.

In the first example, a nondeterministic PDA is constructed for the language

$$\Sigma = \{a, b\}, \ L = \{a^n b^m \mid m > 0, m \le n \le 3m\}.$$

From the menu for FLAP (Figure 1), one selects the type of automaton to construct. Figure 2 shows a constructed PDA for this language. The input string *aaaaaaaabbb* has been entered in the input box, and acceptance by final state has been selected.

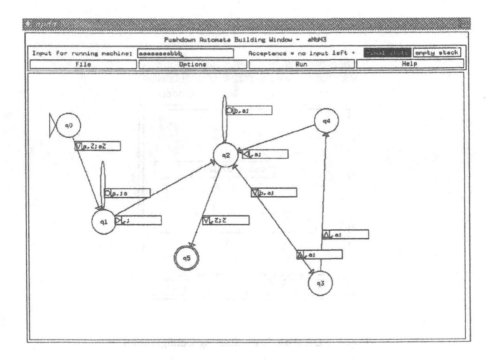

Fig. 2. NPDA for $\{a^n b^m \mid n > 0, n \le m \le 3n\}$

The fast simulation for this PDA and input string results in acceptance, as shown in Figure 3. The total number of nodes in this configuration tree is 75, with the path length from the root (or starting configuration) to an acceptance configuration (final state) of 18. Selecting *Yes* displays an animation of configurations in the acceptance path, starting with the start state (Figure 4) and ending in the final state (see Figure 5). The top rectangle in each figure shows the remaining symbols to be processed in the input string, while the bottom rectangle shows the stack (with the top of the stack on the left).

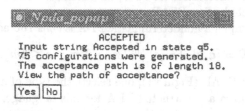

Fig. 3. NPDA Acceptance Information

Fig. 4. PDA Path Trace Starting Configuration

Alternatively, the automaton may be run step-by-step. In this case, a Run Window appears, showing the starting configuration. As the user steps through the simulation, all current configurations (up to 12) are shown. Figure 6 shows a snapshot of the run of the PDA in Figure 2, several steps into the trace (at this point there are 6 possible configurations).

Additional Examples. Below are a list of languages for which automata have been built using FLAP. FLAP is an instructional tool and is not meant to design

Fig. 5. PDA Path Trace Acceptance Configuration

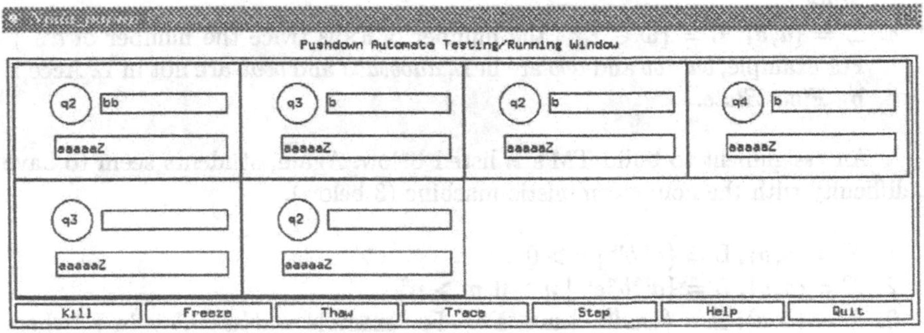

Fig. 6. PDA Running Window

automata with a large number of states. There is a limit to what can be physically drawn in the drawing window, plus it would be too tedious to construct large automata. The lists of languages below were given as lab assignments or homework assignments in CPS 140.

An assignment to build FA is listed below. The examples of strings either in language L or not in L given with each L encourage students to test their machines. For this assignment, the fourth one is really too large to use with FLAP.

1. $\Sigma = \{0,1,2\}$, $L = \{w \in \Sigma^* \mid w$ is a nonnegative integer base 3 $\}$. For example, 0, 1002, and 221 are in L, but 000, and 020 are not in L.
2. $\Sigma = \{a,b\}$, $L = \{w \in \Sigma^* \mid w$ has an even number of a's and an odd number

of b's }. For example, *aabaa* and *bbaba* are in L, *aabbbaba* and *bbab* are not in L.

3. $\Sigma = \{a, b\}$, L = $\{w \in \Sigma^* \mid |w| \bmod 3 = 1\}$

4. $\Sigma = \{a, b\}$, L = $\{w \in \Sigma^* \mid |w| >= 4$ and every substring of four symbols has at least 2 b's}. For example, *aabbaba* is in L (every substring of 4 consecutive symbols has 2 b's) and *aaab* is not in L. The string *ababaa* is not in L because the substring *abaa* does not contain 2 b's.

An assignment to build PDA's is listed below. One can select to accept by either final state or empty stack. Students seem to have difficulty with the nondeterministic machines (2 and 3 below).

1. $\Sigma = \{a, b\}$, L = $\{a^n b^{3n} \mid n > 0\}$. For example, *abbb* and *aabbbbbb* are in L. Accept by *Final State*.

2. $\Sigma = \{a, b\}$, L = $\{w \in \Sigma^* \mid w = w^R$ and w is of odd length }. For example, *bbabb* and *ababa* are in L, *aabbaa* and *bbbab* are not in L. Accept by *Empty Stack*.

3. $\Sigma = \{a, b\}$, L = $\{a^n b^m \mid m > 0, m \leq n \leq 3m\}$. For example, *aaabb* and *aaaaabb* are in L, and *aabbb* and *aaaaaaabb* are not in L. Accept by *Final State*.

4. $\Sigma = \{a, b\}$, L = $\{w \in \Sigma^* \mid$ the number of b's is twice the number of a's }. For example, *bababb* and *abb* are in L, *aabbbaba* and *bbab* are not in L. Accept by *Final State*.

An assignment to build TM's is listed below. Again, students seem to have difficulty with the nondeterministic machine (3 below).

1. $\Sigma = \{a, b\}$, L = $\{a^n b^n \mid n > 0\}$.

2. $\Sigma = \{a, b\}$, L = $\{a^m b^n c^n \mid n > 0, m > n\}$.

3. $\Sigma = \{a, b\}$, L = $\{w w^R w \mid w \in \Sigma^*\}$. For example, *aabbaaaab* is in L, where $w = aab$.

An assignment to build two-tape TM's included the following language.

1. $\Sigma = \{a, b\}$, L = $\{a^m b^n c^n \mid n > 0, m > n\}$.

2.4 Examples Using FLAP in Lecture

We use FLAP projected onto a screen during lectures to solve problems and to aid in the discussion of proofs. As part of the lecture, students are given a few minutes to construct an automaton independently, after which the instructor, with input from the class, uses FLAP to construct the automaton on the screen. In this manner, students actively discover the thinking process in designing an automaton, while also learning how to use FLAP.

We examine proofs of converting three types of grammars into an NPDA: a context-free grammar in Greibach normal form, an LL(1) grammar, and an

LR(1) grammar. For each of these, we work through a simple example constructing the corresponding NPDA and testing input strings. For the latter two, we proceed to talk in depth about the LL(1) and LR(1) parsing process. Since the PDA created is nondeterministic, we can run these machines on input strings in the slow mode and determine the lookahead (or rather, which configuration to try next). This method aids in understanding both parsing processes.

3 LLparse and LRparse

In this section we give an overview of LLparse and LRparse [4] and give an example using LRparse.

3.1 Overview of LLparse and LRparse

LLparse and LRparse are visual and interactive instructional tools for constructing LL(1) and LR(1) parse tables [1] from appropriate grammars, and for using these constructed tables to parse strings. Naturally, there are size limitations due to what can visually be displayed. These tools allow reasonable size examples for experimenting with and understanding these methods.

For the most part, the two tools have a common interface. Both tools consist of a series of windows representing the steps in building a parse table. The user cannot proceed to the next window until the current step is correctly completed.

For both tools, the first three windows encountered are the same, except that in LLparse, an LL(1) grammar with at most fifteen rules is entered in the first window, while in LRparse, an LR(1) grammar is entered. After such a grammar is entered successfully, a second window appears, displaying a table of blank FIRST sets of appropriate strings that need to be filled in. Upon correct entry of these sets, a similar window appears for the FOLLOW sets of variables.

After this point in LLparse, a fourth window appears containing an empty LL(1) parse table with the correct number of labeled columns and rows. Upon successful entry of the parse table, a parsing window appears, which allows the user to enter an input string and start an animation visualizing the step-by-step parsing of this string, showing the current stack contents and explaining which entries in the table are being used.

The fourth window appearing in LRparse, a modified version of the DFA window in FLAP, requires the entry of a DFA representing the states in the parsing process. By using the mouse, the user can graphically draw states and labeled arcs representing a DFA. The number of states in the DFA is limited to at most twenty-five. The item sets can be generated for each state by clicking on the state and entering the set of items in a small window corresponding to the state. Upon successful creation of the correct DFA and item sets, a window representing the LR(1) parse table appears with the correct number of labeled rows and columns. Once the table is successfully filled in, the final parsing window appears, which allows example strings to be visually parsed using the constructed LR(1) parse table.

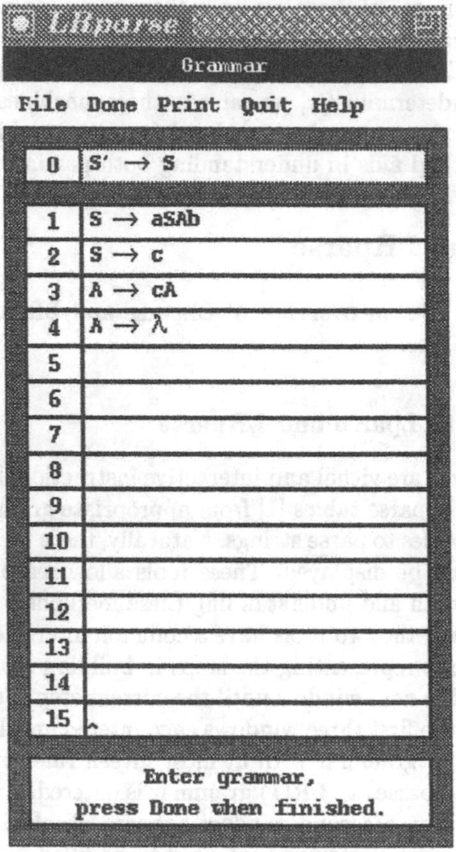

Fig. 7. Enter Grammar in LRparse

All the windows contain a subset of the following buttons. The *DONE* button announces that a user has finished typing input in a window. At this point, the user is informed whether the entries are correct, and if not correct, the incorrect entries are highlighted. Since the correctness of the DFA is more complicated, only one error at a time is highlighted. The *SHOW* button automatically fills in the window with the correct entries. In the DFA window, it automatically constructs the DFA. This is useful for frustrated students. The *PRINT* button prints to paper or to a file all of the work done, up to the current window. The *HELP* button provides online help, the *QUIT* button allows the user to exit, and the *RETURN* button allows the user to return to the initial window to start over.

In both tools, although the parse table is immediately calculated after the grammar is entered, it is not shown. Thus, if it is determined that the grammar

is not an LL(1) or LR(1) grammar, the user is informed and given the option
to continue or change the grammar. The user can continue with the incorrect
grammar up to the calculation of the parse table, at which point some position
in the table will have multiple entries.

3.2 Example of LRparse

We present an example using LRparse to construct an LRparse table for the
following LR(1) grammar.

$$S \rightarrow aSAb \mid c$$
$$A \rightarrow cA \mid \lambda$$

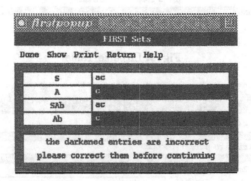

Fig. 8. First Sets in LRparse with a Mistake

Upon starting LRparse, the grammar window in Figure 7 initially would
appear empty, and the user would type in the grammar. The arrow and λ are
typed using "CTRL ." (note that the ">" is on the same key as the ".") and
"CTRL l", as explained in the provided online help. After selecting **Done**, the
FIRST sets window appears. In Figure 8 the user has been informed that the
FIRST sets for A and Ab are incorrect. (The FIRST set for A should be "cλ",
and the FIRST set for Ab should be "cb"). Upon correcting the mistake and
selecting **Done**, a similar FOLLOW set window appears for the user to fill in.
Next, a blank Build window appears that is similar to the DFA part of FLAP.
The user can construct the corresponding DFA that models the symbols on
top of the stack. The DFA differs from the DFA in FLAP in that an item set
must be entered for each state. A small window pops up after selecting a state,
and one enters the marked rules for that state. This window can stay up or be
hidden. After completing the correct DFA, selecting **Done** brings up the parse
table window. Figure 9 shows the parse table for the grammar in Figure 7. By
selecting row 3, the item sets for state q3 can be displayed and removed when
desired.

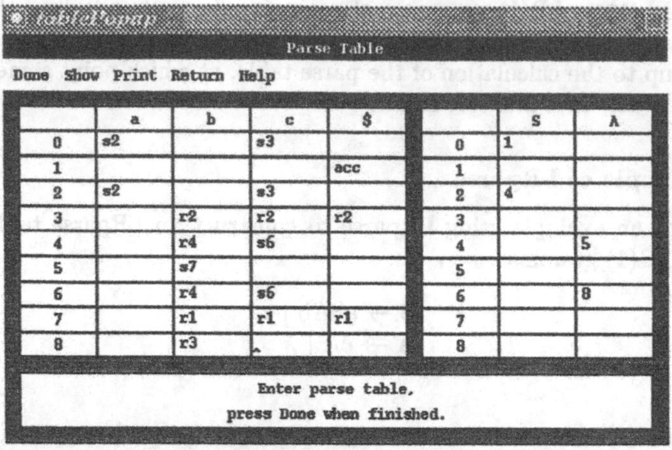

Fig. 9. LRparse Table

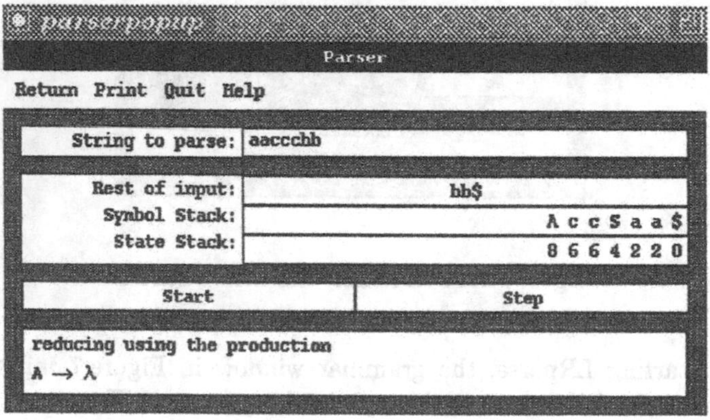

Fig. 10. LRparse Parsing Example

When the parse table is correct, the final window displayed is the parsing window. Figure 10 shows part of the trace of the string *aacccbb*. Informative messages are displayed at the bottom of the window, describing what type of operation (reduce, shift, accept, or error) has just been performed.

3.3 Using LLparse and LRparse in a Course

We use LLparse and LRparse during lectures to increase the interaction in the classroom. For example, when describing how to calculate FIRST sets, students initially work through an example at their desk for several minutes, after which the instructor tries out a student's solution using the tool. If there is a mistake, it will soon be evident. After creating parse tables, the instructor steps through the parsing of a string suggested by the students.

Outside of lectures, students use LLparse and LRparse in a homework assignment. They also use these tools heavily to study for exams, thus ensuring that they understand the algorithms for constructing the parse tables.

4 Interpreter Programming Assignment

Writing an interpreter is beneficial in many ways. Students can observe how material from the automata theory course, as well as from their previous data structures course, applies to the real world. They also gain additional programming experience. In CPS 140, students write an LR(1) parser for a new programming language in three phases: the scanner, the parser, and the syntax tree. In the third phase, the tool Xtango [10] is used to animate the execution of a program in the new language.

Each phase is built from scratch. The new programming language is very simple, allowing students to focus on how FA's and PDA's are used in the parsing process. Tools such as lex and yacc are not used, although they are discussed at the end of the semester so that students are aware of them if they are creating another language.

In the first phase, students write a scanner from scratch and set up a symbol table (or hash table). Their output is a list of tokens and their type. In the second phase, students focus on the LR(1) parsing process. They are required to create a stack, process operations (shift, reduce, accept and error), and print out the rules found (in reverse order) if a program is syntactically correct. The parse table is given to students (in a file), since the parse table can be quite large (in the example language below it is 41 rows and 24 columns). In the third phase, a syntax tree is created, enabling the program to be interpreted by walking through the tree.

4.1 An Example Language and the Use of Xtango

A different simple programming language is created each semester. In the spring 1996 semester, students wrote an interpreter for the RIMLAN programming language, a simple language for moving robots around a room that contains six types of statements.

Statement	Meaning
begin *i j stmts* halt	program definition - defines starting room of height *i* and width *j*
robot *v a b* ;	draw a robot *v* at position (a, b)
obstacle *a b* ;	draw an obstacle at position (a, b)
add *a* to *v* ;	add statement
move *v d a* ;	move the robot *v*, *a* spaces in direction *d*
v = a ;	an assignment statement
do *stmts* until *a > b* ;	Execute *stmts*, if $a \leq b$ then repeat

where *v* is a variable, *a* and *b* are either variables or integers, *i* and *j* are integers, *d* is a direction (north, south, east or west) and *stmts* represents 1 or more valid statements.

Below is a sample RIMLAN program that creates a robot named bob and two obstacles, and then makes bob circle the perimeter of the rectangle formed by the two obstacles ten times.

```
begin 30 40
    robot bob 2 4 ;
    obstacle 4 5 ;
    obstacle 7 12 ;
    horizontal = 4 ;
    vertical = 8 ;
    move bob east 6 ;
    j = 1 ;
    do
        move bob north vertical ;
        move bob west horizontal ;
        move bob south vertical ;
        move bob east horizontal ;
        add 1 to j ;
    until j > 10 ;
halt
```

Visualizing the running of RIMLAN programs helps to determine whether robots have crashed into obstacles or other robots. We use the tool Xtango to create simple animations. In particular, we use the Xtango animator (an interpreter) which allows one to create simple objects such as rectangles, circles, points and lines. Each created object has an associated tag, which can be referred to in order to move the object. This makes it very simple to create obstacles (squares) and robots (circles) and then move the robots.

5 Evaluation

We have been teaching automata theory for seven years (five years at Rensselaer Polytechnic Institute and the past two years at Duke University). During the first two years, the programming assignment was part of the course, but no tools were used and the programming assignment did not have an animation component. Since then, the tools have been integrated into lectures and assignments, and this year FLAP was used for the first time in a separate computer lab.

Although FLAP has undergone many changes and improvements over the years, students have been enthusiastic about using it to design and run automata. In CPS 140 (spring 1996), FLAP was used by students in three lab periods: The first to design FA; the second to design PDA's; and the third to design TM's and two-tape TM's. After the first lab, about half the students still preferred to write the FA on paper first so they could write comments beside states. However, the majority of students stated in evaluations that they found FLAP very easy to use and were impressed with it, and all students found it beneficial to be able to run the programs.

Comments after the first lab using FLAP include the following.

- "The flap program was easy to use. It took all of two seconds to learn."
- "The benefit of using FLAP is that you can easily test your design. This benefit outweighs the hassle of having to use a program to design the DFA. By using this program you can be sure your tests are conducted correctly and they don't take any time to run."
- "I found myself creating the basic structure for most on paper, and then testing them thoroughly via FLAP. FLAP was a lot faster (and more accurate) at testing the FA's than pencil and paper."
- "All in all, I was actually impressed with FLAP, and I found it to be easier to think with than the pencil-and-paper method, and easier to understand the finished product than with hand-drawn FA."

Comments after the third lab using FLAP include the following.

- "Flap was very easy to use. I think the fact that the Turing Machines are more complicated to test on paper makes flap a very good tool."
- "I still used paper to lay the groundwork for my TM's. Then, I transferred them to Flap and modified them when they didn't work. So, it is good that Flap is there to catch your errors."
- "Creating Turing machines was actually kind of fun. It required a little more thought. FLAP definitely came in handy here to test all the different cases that you could get. Made things a lot more efficient."
- "FLAP is getting to be more worthwhile as the complexity of the things we are modelling increases. The trace function is incredibly useful, especially the two taped TM."
- "I now find it much easier to create FA on FLAP than by hand."
- "I like FLAP more as I get used to it. I find it to be very valuable when designing complex machines. It makes things much easier to test."

6 Conclusion and Future Work

We have transformed the Formal Languages and Automata Theory course from a course with a slow and small amount of feedback into a course using visual and interactive tools, increasing the amount and quickness of the feedback. The tool FLAP allows one to create and simulate several types of automata, and the tools LLparse and LRparse allow one to create parse tables and parse arbitrary input strings. Together with a three part programming assignment writing an LR(1) parser, these tools greatly enhance the amount of hands-on work in this course. Students are very positive about these tools, since they enjoy being able to test out their automata designs using FLAP, and check their construction of a parse table using LLparse or LRparse.

At this point, we have not converted all assignments into this format. We still give some written homework assignments, but we are currently investigating how to create tools for all of our assignments. Current work on these tools include creating a Java version of FLAP called JFLAP, and adding parse trees to LLparse and LRparse. We are also investigating tools for experimenting with grammars. These tools are available via anonymous ftp from our web site
`http://www.cs.duke.edu/~rodger`

7 Acknowledgments

Many students from Rensselaer and Duke have worked on designing and developing FLAP, LLparse, and LRparse, including Dan Caugherty, Mark LoSacco, Greg Badros, Magda Procopiuc, Octavian Procopiuc, Mike James, Steve Blythe, Ugur Dogrosou, and Edwin Tsang. Thanks to Caroline Sori for comments on the paper.

References

1. Aho, A., Sethi, R., Ullman, J.: Compilers: Principles, Techniques, and Tools. Addison-Wesley (1986)
2. Badre, A., Lewis, C., Stasko, J.: Do algorithm animations assist learning? An empirical study and analysis. INTERCHI 93 Conference Proceedings: Human Factors in Computing Systems, ACM Press (April 1993) 61–66
3. Badre, A., Lewis, C., Stasko, J.: Empirically Evaluating the Use of Animations to Teach Algorithms. Proceedings of the 1994 IEEE Symposium on Visual Languages (1994) 48–54
4. Blythe, S., James, M., Rodger, S.: LLparse and LRparse: Visual and Interactive Tools for Parsing. Twenty-fifth SIGCSE Technical Symposium on Computer Science Education (1994) 208–212
5. Caugherty, D., Rodger, S. H.: NPDA: A Tool for Visualizing and Simulating Nondeterministic Pushdown Automata. in Computational Support for Discrete Mathematics, DIMACS Series in Discrete Mathematics and Theoretical Computer Science, Vol. 15, N. Dean and G. E. Shannon (ed.), American Mathematical Society (1994) 365–377

6. Lewis, H., Papadimitriou, C.: Elements of the Theory of Computation. Prentice-Hall (1981)
7. Linz, P.: An Introduction to Formal Languages and Automata. D. C. Heath and Company (1990)
8. LoSacco, M., Rodger, S. H.: FLAP: A Tool for Drawing and Simulating Automata, ED-MEDIA 93, World Conference on Educational Multimedia and Hypermedia (1993) 310–317
9. Rodger, S. H.: An Interactive Lecture Approach to Teaching Computer Science, Twenty-sixth SIGCSE Technical Symposium on Computer Science Education (1995) 278–282
10. Stasko, J.: Tango: A Framework and System for Algorithm Animation. IEEE Computer (1990) 27–39

NFA to DFA Transformation for Finite Languages*

Kai Salomaa and Sheng Yu

Department of Computer Science
University of Western Ontario
London, Ontario, Canada N6A 5B7

Abstract. We consider the number of states of a DFA that is equivalent to an n-state NFA accepting a *finite language*. We first give a detailed proof for the case where the finite languages are over a two-letter alphabet. It shows that $O(2^{n/2})$ is the (worst-case) optimal upper-bound on the number of states of a DFA that is equivalent to an n-state NFA accepting a finite language. The main result of this paper is a generalization of the above result. We show that, for any n-state NFA accepting a finite language over an arbitrary k-letter alphabet, $n, k > 1$, there is an equivalent DFA of $O(k^{n/(\log_2 k + 1)})$ states, and show that this bound is optimal in the worst case.

1 Introduction

It is well-known that for each positive integer n, there exists a regular language L such that L is accepted by an n-state NFA and any complete DFA accepting L requires at least 2^n states [3]. However, the same statement is not true if L is required to be finite. In [2], Mandl showed that for each n-state NFA accepting a finite language over *a two-letter alphabet*, there exists an equivalent DFA which has $O(2^{\frac{n}{2}})$ states; more specifically, no more than $2^{\frac{n}{2}+1} - 1$ states if n is even and $3 \cdot 2^{\lfloor \frac{n}{2} \rfloor} - 1$ states if n is odd. In [2], it was also shown that these bounds are optimal in the worst case. However, there have been no corresponding results concerning finite languages over an arbitrary k-letter alphabet, $k \geq 2$. Also, the proofs in [2] for the two-letter alphabet case are rather sketchy.

In this paper, we first give detailed proofs for the two-letter alphabet case. Then, as the main result of this paper, we give the optimal upper-bounds for the general cases of the problem, i.e., for the cases where finite languages are over an arbitrary k-letter alphabet, $k \geq 2$. Specifically, we show that for any n-state NFA accepting a finite languages over a k-letter alphabet, $k \geq 2$, we can construct an equivalent DFA of $O(k^{n/(\log_2 k + 1)})$ states; and we show that for each k-letter alphabet, $k \geq 2$, and for each $n \geq 2$, there exists a finite language L accepted by an n-state NFA such that the number of states of any DFA accepting L is

---
* This research is supported by the Natural Sciences and Engineering Research Council of Canada grants OGP0041630.

$\Omega(k^{n/(\log_2 k+1)})$. One may observe that this bound, as a function of k, approaches to 2^n when k becomes larger.

This paper is organized as follows. In the next section, we introduce the basic notations and definitions and also review some necessary background. In Section 3, we prove preliminary results that are necessary to the proofs in the next two sections. In Section 4, we give detailed proofs for the optimal upper-bounds for finite languages over a two-letter alphabet. In Section 5, we prove the general result, i.e., that for an arbitrary k-letter alphabet Σ, $k > 1$, the optimal upper-bound on the number of states of a DFA that is equivalent to a n-state NFA accepting a finite language over Σ is $O(k^{n/(\log_2 k+1)})$.

2   Basic notations and definitions

A *deterministic finite automaton* (DFA) A is a quintuple $(Q, \Sigma, \delta, s, F)$, where

Q is the finite set of states;

Σ is the input alphabet;

$\delta : Q \times \Sigma \to Q$ is the state transition function;

$s \in Q$ is the starting state; and

$F \subseteq Q$ is the set of final states.

Note that the transition function δ is not necessarily a total function, i.e., it may not be defined for every pair of state and input symbol in $Q \times \Sigma$. If δ is a total function, we call A a *complete DFA*.

For convenience, we define $\delta^+ : Q \times \Sigma^+ \to Q$ recursively by

(1) $\delta^+(q, a) = \delta(q, a)$ and

(2) $\delta^+(q, xa) = \delta(\delta^+(q, x), a)$

for $q \in Q$, $a \in \Sigma$, and $x \in \Sigma^+$.

A *nondeterministic finite automaton* A is a quintuple $(Q, \Sigma, \delta, s, F)$ where Q, Σ, s, and F are defined exactly the same way as for a DFA, and $\delta : Q \times \Sigma \to 2^Q$ is the transition function, where 2^Q denotes the power set of Q.

Similarly, we define $\delta^+ : Q \times \Sigma^+ \to 2^Q$ for an NFA A by

(1) $\delta^+(q, a) = \delta(q, a)$ and

(2) $\delta^+(q, xa) = \cup_{p \in \delta^+(q,x)} \delta(p, a)$.

A DFA or an NFA such that every state is reachable from the starting state and reaches a final state is called a *reduced* DFA or NFA, respectively.

The language accepted by a finite automaton A, either an NFA or a DFA, is denoted $L(A)$. We say that two finite automata A and B are equivalent if they accept exactly the same language, i.e., $L(A) = L(B)$.

Note that according to the above NFA definition, an NFA does not have any λ-transitions. We call a nondeterministic finite automaton that has λ-transitions a λ-NFA. It has been shown in [9] that each λ-NFA is equivalent to an NFA with the same number of states. Therefore, the subsequent results on the number of states of NFA apply to λ-NFA as well.

The standard algorithm that transforms an NFA to an equivalent DFA is called the subset construction algorithm. Let $A = (Q, \Sigma, \delta, s, F)$ be an arbitrary

NFA. Following the algorithm, an equivalent DFA $A' = (Q', \Sigma, \delta', s', F')$ is constructed such that each state in Q' is a subset of Q. The transition function $\delta' : Q' \times \Sigma \to Q'$ is defined by

$$\delta'(P_1, a) = P_2 \text{ if } P_2 = \{q \in Q \mid \text{there exists } p \in P_1 \text{ such that } q \in \delta(p, a) \}.$$

Note that we construct for Q' only those subsets of Q that can be reached from the starting state $s' = \{s\}$.

For regular languages in general, it is possible that Q' consists necessarily of all the 2^n subsets of Q, assuming that $|Q| = n$. For example, any complete DFA that is equivalent to the n-state NFA shown in Figure 1 has at least 2^n states [3].

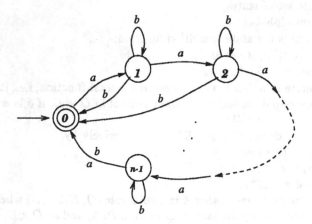

Fig. 1. An n-state NFA such that any equivalent DFA has at least 2^n states

For a language $L \subseteq \Sigma^*$, the binary relation $\equiv_L$ on words in Σ^* is defined by, for $x, y \in \Sigma^*$, $x \equiv_L y$ iff $xz \in L \Leftrightarrow yz \in L$ for all $z \in \Sigma^*$. It is clear that $\equiv_L$ is an equivalence relation. The following is a well-known result, see, e.g., [4, 5, 1].

Theorem 1 *Let $L \in \Sigma^*$ be a regular language. Then there is a bijection between the equivalence classes of $\equiv_L$ and the states of a minimal complete DFA A that accepts L. Thus, the number of equivalence classes of $\equiv_L$ is equal to the number of states of A.*

Our notations follow those used in [1] in general. Note that, for a word $x \in \Sigma^*$, we denote by $|x|$ the length of x. For a set S, we use the same notation, $|S|$, to denote the number of elements in S if this does not cause any confusion.

3 Preliminary results

In this section, we prove several technical results which are essential to the proofs of the results in the next two sections.

Let $A = (Q, \Sigma, \delta, s, F)$ be an n-state NFA accepting a finite language over a k-letter alphabet, $k > 1$. We assume that A is reduced. Then it is clear that A does not contain any cycles in its state-transition, i.e., for any $p, q \in Q$, if $q \in \delta^+(p, x)$ for some $x \in \Sigma^+$, then $p \neq q$.

We construct a DFA $A' = (Q', \Sigma, \delta', s', F')$ using the standard subset construction algorithm and assume that Q' contains only those states that are reachable from s'. For convenience, we also assume that A' is complete.

Lemma 1 *Let $P = \{P_1, \ldots, P_m\}$ and $R = \{R_1, \ldots, R_n\}$ be two subsets of Q' such that $P \cap R = \emptyset$ and each R_j in R is reachable from some state P_i in P, i.e., $R_j \in \delta'^+(P_i, x)$ for some $P_i \in P$ and $x \in \Sigma^+$. (Note that $P_1, \ldots, P_m, R_1, \ldots, R_n$ are all subsets of Q). Then there is $p \in Q$ such that*

$$p \in \bigcup_{i=1}^{m} P_i \text{ but } p \notin \bigcup_{j=1}^{n} R_j.$$

Proof. We prove this by contradiction. Assume that there does not exist such p, i.e.,

$$\bigcup_{i=1}^{m} P_i \subseteq \bigcup_{j=1}^{n} R_j. \tag{1}$$

Note that since for each $R_j \in R$, there are $P_i \in P$ and $x \in \Sigma^+$ such that $R_j \in \delta'^+(P_i, x)$, it holds that for each $r \in \bigcup_{j=1}^{n} R_j$, there are $p \in \bigcup_{i=1}^{m} P_i$ and $x \in \Sigma^+$ such that $r \in \delta^+(p, x)$.

Let t be the number of distinct states (of Q) in $\bigcup_{i=1}^{m} P_i$. Consider an arbitrary state p_1 in $\bigcup_{i=1}^{m} P_i$. From the inclusion (1), it follows that $p_1 \in \bigcup_{j=1}^{n} R_j$. Then p_1 is reachable from some state in $\bigcup_{i=1}^{m} P_i$. Denote this state by p_2, that is, $p_1 \in \delta^+(p_2, x_1)$ for some $x_1 \in \Sigma^+$. If $p_1 = p_2$, then there is a cycle in the transition diagram of A. This contradicts the fact that A accepts a finite language. Thus, we assume that $p_1 \neq p_2$. Again, p_2 is reachable from some state $p_3 \in \bigcup_{i=1}^{m} P_i$, i.e., $p_2 \in \delta^+(p_3, x_2)$ for some $x_2 \in \Sigma^+$. If $p_3 = p_1$ or $p_3 = p_2$, then we obtain a contradiction. So, we assume that $p_3 \neq p_1$ and $p_3 \neq p_2$. Continue such arguments until we see that p_t has to be reached from some state p_i, $1 \leq i \leq t$. That is $p_t \in \delta^+(p_i, x_t)$ for some $x_t \in \Sigma^+$. Since $p_t \in \delta^+(p_i, x_t)$, $p_i \in \delta^+(p_{i+1}, x_i)$, $\ldots$, $p_{t-1} \in \delta^+(p_t, x_{t-1})$, we have a cycle $p_t \in \delta^+(p_t, x_{t-1}x_{t-2} \cdots x_i x_t)$. This is a contradiction. Therefore, the assumption that there does not exist a state $p \in Q$ such that $p \in \bigcup_{i=1}^{m} P_i - \bigcup_{j=1}^{n} R_j$ is false. Thus, we have proved the lemma. $\square$

Definition 1 *We define levels for the states in Q' recursively as follows:*

(1) The starting state s' is said to be of level 0.

(2) Let $P_1 \in Q'$ be of level i, $i \geq 0$. If $P_2 = \delta'(P_1, a)$ and P_2 is not of level j for any $j \leq i$, then P_2 is said to be of level $i + 1$.

In other words, the level of a state P in Q' is the length of the shortest word from the starting state s' to P.

Clearly, the highest possible level of any state in Q' except the empty subset $\emptyset$ of Q is $n-1$, since no words in L are more than $n-1$ characters long. The only state of Q' that can be of level n is the empty subset $\emptyset$ of Q.

Definition 2 *For each i, $0 \leq i \leq n$, define Q'_i to be the set of all states of level i in Q'.*

Lemma 2 *For each i, $0 \leq i \leq n-1$, there is $q \in Q$ such that $q \in P_s$ for some state $P_s \in Q'_i$ but $q \notin R_t$ for any $R_t \in Q'_j$ for $j > i$.*

Proof. This lemma follows directly from Lemma 1: Consider Q'_i to be P of Lemma 1 and $\bigcup_{j>i} Q'_j$ to be R of Lemma 1. $\square$

Lemma 3 *For each i, $0 \leq i \leq n$,*

$$\left| \bigcup_{j \geq i} Q'_j \right| \leq 2^{n-i}.$$

Proof. By Lemma 2, all states in $\bigcup_{j \geq i} Q'_j$ are subsets of a set $S_i \subseteq Q$ of no more than $n-i$ states. Then the number of all distinct subsets of S_i is 2^{n-1}. Thus, the lemma holds. $\square$

Corollary 1 *For each i, $0 \leq i \leq n$,*

$$|Q'_i| \leq 2^{n-i}.$$

4 Finite languages over a two-letter alphabet

Here and in the next section, we will consider the NFA to DFA transdormation for finite languages over two-letter alphabets and general k-letter alphabets, $k \geq 2$, respectively. The one-letter alphabet case is rather trivial. However, for completeness, we first state the following result for the one-letter alphabet case.

Theorem 2 *Let L be a finite language over a one-letter alphabet Σ and m be the length of the longest word in L, $m \geq 0$. Then the minimum complete DFA accepting L has exactly $m+2$ states.*

Therefore, each n-state NFA that accepts a finite language over a one-letter alphabet is equivalent to a complete DFA with at most $n+1$ states.

Theorem 3 *Let $\Sigma = \{a, b\}$ and $L \subseteq \Sigma^*$ be an arbitrary finite language accepted by an n-state NFA $A = (Q, \Sigma, \delta, s, F)$, $n \geq 2$. Then there is a complete DFA $A' = (Q', \Sigma, \delta', s', F')$ of no more than $2^{\lceil \frac{n}{2} \rceil + 1} - 1$ states such that $L = L(A')$. More precisely, A' has at most $2^{\frac{n}{2}+1} - 1$ states if n is even and $3 \cdot 2^{\lfloor \frac{n}{2} \rfloor} - 1$ states if n is odd.*

Proof. Since A' is deterministic, we have $|Q'_i| \leq 2^i$, $0 \leq i \leq n$. By Corollary 1, we have $|Q'_i| \leq 2^{n-i}$. So, $|Q'_i| \leq min(2^i, 2^{n-i})$.

If n is even, then clearly $|Q'_i|$ reaches maximum when $2^i = 2^{n-i}$, i.e., $i = \frac{n}{2}$. Then, it is not difficult to show that the maximum possible number of states in Q' is

$$1 + 2 + 2^2 + \ldots + 2^{\frac{n}{2}} = 2^{\frac{n}{2}+1} - 1.$$

Note that by Lemma 3, the total number of all possible states of levels $\frac{n}{2}$, $\frac{n}{2}+1$, ..., n is at most $2^{\frac{n}{2}}$.

If n is odd, the maximal possible value for $|Q'_i|$ is $2^{\lfloor \frac{n}{2} \rfloor}$, which can be reached at both $i = \lfloor \frac{n}{2} \rfloor$ and $i = \lceil \frac{n}{2} \rceil$. Then, it can be shown that the maximum possible number of states in Q' is

$$1 + 2 + \ldots + 2^{\lfloor \frac{n}{2} \rfloor} + 2^{\lfloor \frac{n}{2} \rfloor} = 3 \cdot 2^{\lfloor \frac{n}{2} \rfloor} - 1.$$

Note again that $|\cup_{j \geq \lceil \frac{n}{2} \rceil} Q_j| \leq 2^{\lfloor \frac{n}{2} \rfloor}$. $\square$

We now prove that the upper-bounds given above can be reached. In other words, the upper-bounds are optimal.

Theorem 4 *For each integer $n > 1$, there exists a finite language $L \subseteq \{a,b\}^*$ such that L is accepted by an n-state NFA and any complete DFA accepting L has at least $2^{\frac{n}{2}+1} - 1$ states if n is even or $3 \cdot 2^{\lfloor \frac{n}{2} \rfloor} - 1$ states if n is odd.*

Proof. Let $\Sigma = \{a, b\}$. Consider the language

$$L_n = \{w \in \Sigma^* \mid w = uav \text{ such that } |w| < n \text{ \& } |v| = \lfloor \frac{n}{2} \rfloor - 1\},$$

for $n \geq 2$. Intuitively, L_n is the set of all words w such that $\lfloor \frac{n}{2} \rfloor \leq |w| < n$ and their $\lfloor \frac{n}{2} \rfloor$th letter from the right is an a. It is clear that L_n is accepted by an n-state NFA.

There are two cases to consider: (1) n is even and (2) n is odd. In the first case, let $n = 2m$. We now count the number of distinct equivalence classes of the relation $\equiv_{L_n}$. Let $x, y \in \Sigma^*$, $|x| = i$ and $|y| = j$, $0 \leq i, j \leq m$. It is clear that $x \not\equiv_{L_n} y$ if $i \neq j$ (assuming that $i < j$) since $xa^{n-i} \in L_n$ but $ya^{n-i} \notin L_n$. For $|x| = |y| = i$, $1 \leq i \leq m$, we prove that $x \not\equiv_{L_n} y$ if $x \neq y$. Let x and y differ at their t^{th} letter from the left. Without loss of generality we can assume that the tth letter of x is an a and the tth letter of y is a b. Let $z = a^{m-i+t-1}$. Then $xz \in L_n$ but $yz \notin L_n$. However, no word $x \in \Sigma^*$ such that $|x| > m$ represents a new equivalence class. Consider $x \in \Sigma^*$ such that $m < |x| < n$. Let $d = n - |x|$ and $x = uvw$ such that $|vw| = m$ and $|w| = m - d$. Then it can be shown that $x \equiv_{L_n} vb^{m-d}$. All words $x \in \Sigma^*$ such that $|x| \geq n$ satisfy $x \equiv_{L_n} b^m$. Therefore, there are altogether

$$1 + 2 + 2^2 + \ldots + 2^m = 2^{m+1} - 1$$

distinct classes of $\equiv_{L_n}$.

In the second case, i.e., when n is odd, let $m = \lfloor n/2 \rfloor$. Then $n = 2m + 1$. We can show that every word w such that $|w| \leq m$ represents a distinct equivalence class of $\equiv_{L_n}$ using a similar argument as above. However, for $w \in \Sigma^{m+1}$, this case differs from the previous case. In this case, any word x such that $|x| = m + 1$ satisfies $x \not\equiv_{L_n} y$ for any word y such that $|y| < m + 1$ since $ya^m \in L_n$ but $xa^m \notin L_n$. Note that two words $x, y \in \Sigma^*$ such that $|x| = |y| = m + 1$ belong to the same equivalence class if they have exactly the same suffix of m letters. Similarly to the previous case, any $x \in \Sigma^*$ such that $|x| > m + 1$ does not represent a new class. If $|x| \geq n$, then $x \equiv_{L_n} b^m$. If $m + 1 < |x| < n$ and $x = uvw$ such that $|vw| = m + 1$ and $|w| = m - n + |x|$, then $x \equiv_{L_n} vb^{|w|}$. So there are in total

$$1 + 2 + 2^2 + \ldots + 2^m + 2^m = 3 \cdot 2^m - 1$$

distinct equivalence classes of $\equiv_{L_n}$. $\square$

Corollary 2 *The upper-bounds given in Theorem 3 are optimal for all $n \geq 2$.*

5 Finite languages over an arbitrary alphabet

In this section, we consider NFA to DFA transformation for finite languages over an arbitrary k-letter alphabet, $k \geq 2$. Note that, in [2], the section title for the two-letter alphabet results is "Finite languages over arbitrary alphabets". However, the results stated there concern only finite languages over a two-letter alphabet.

Theorem 5 *Let Σ be a k-letter alphabet, $k \geq 2$, and $L \subseteq \Sigma^*$ be an arbitrary finite language accepted by an n-state NFA $A = (Q, \Sigma, \delta, s, F)$. Then we can construct a DFA $A' = (Q', \Sigma, \delta', s', F')$ of at most $(k^{\lceil \frac{n}{\log_2 k + 1} \rceil + 1} - 1)/(k - 1)$ states such that $L(A') = L$.*

Proof. We construct the DFA A' using the standard subset construction method and assume that Q' contains only those states that are reachable from the new starting state s'. Then it is clear that the i-th level of Q', denoted Q'_i, contains at most k^i states. By also Lemma 3, we have

$$|Q'_i| \leq min(k^i, 2^{n-i}).$$

Let $i_0 = \frac{n}{\log_2 k + 1}$. If i_0 is an integer, then clearly $|Q'_i|$ reaches maximum when $k^i = 2^{n-i}$, i.e., $i = i_0$. Then it can be shown using a similar argument as in the proof of Theorem 3 that the total number of states in Q' of all levels does not exceed

$$1 + k + k^2 + \ldots + k^{i_0} = (k^{i_0+1} - 1)/(k - 1) = (k^{\frac{n}{\log_2 k + 1} + 1} - 1)/(k - 1).$$

If i_0 is not an integer, let $i = \lfloor i_0 \rfloor$. Then we have $k^i < 2^{n-i}$ and $k^{i+1} > 2^{n-i-1}$. The total number of states is no more than

$$1 + k + k^2 + \ldots + k^{\lceil i_0 \rceil} = (k^{\lceil \frac{n}{\log_2 k + 1} \rceil + 1} - 1)/(k - 1).$$

Thus, the theorem holds. □

If the size of the alphabet is $k = 2^t$, for some integer $t \geq 1$, then for any n-state NFA accepting a finite language over the alphabet, there is an equivalent DFA of no more than $(2^{t\lceil \frac{n}{t+1}\rceil+1} - 1)/(k-1)$ states. This is $O(2^{\frac{n}{2}})$ when $k = 2$, $O(2^{\frac{2n}{3}})$ when $k = 4$, and $O(2^{\frac{3n}{4}})$ when $k = 8$. One may observe that the bound approaches to $O(2^n)$ when k becomes larger.

Now we show that those bounds can be reached, i.e., they are optimal.

Theorem 6 *Let k be an arbitrary integer such that $k \geq 2$. Then for any integer $n \geq 2$, there exists a finite language L_n such that L_n is accepted by an n-state NFA and any DFA accepting L_n needs at least $(k^{\lfloor \frac{n}{\log_2 k+1}\rfloor+1} - 1)/(k-1)$ states.*

Proof. Let Σ be a k-letter alphabet and let $t = \lceil \log_2 k \rceil$. We encode each letter in Σ with a distinct t-digit binary sequence. (One letter is encoded by all zeros.) Denote by $E(a)$ the encoding of $a \in \Sigma$ and by $E_i(a)$ the ith digit of $E(a)$ from the left. Define, for each i, $1 \leq i \leq t$,

$$S_i = \{a \in \Sigma \mid E_i(a) = 1\}.$$

Let $m = \lfloor n/(t+1) \rfloor$. We define

$$L_n' = \{w = x_t a_t x_{t-1} \ldots a_1 x_0 \mid$$
$$a_1 \in S_1, \ldots, a_t \in S_t, \ x_1, \ldots, x_t \in \Sigma^{m-1}, \text{ and } |w| = n-1\}.$$

Let L_n be the set of suffixes of length at least m of words in L_n'.

It is easy to see that the language L_n' is accepted by a DFA A' consisting of a "chain" of n states. The DFA A' verifies that the length of the input is $n-1$ and that the symbols occur in the position im are from S_i, $i = 1, \ldots, t$, An NFA A accepting L_n is obtained from A' by adding λ-transitions from the initial state to all except the last m states.

Now we show that the number of equivalence classes of $\equiv_{L_n}$ is at least $(k^{m+1} - 1)/(k-1)$.

Consider all words w in Σ^* such that $|w| \leq m$. Let $w_1, w_2 \in \Sigma^*$ and $|w_1|, |w_2| \leq m$. If $|w_1| \neq |w_2|$, then clearly $w_1 \not\equiv_{L_n} w_2$. If $|w_1| = |w_2|$ but $w_1 \neq w_2$, then there exist $a, b \in \Sigma$ and $a \neq b$ such that $w_1 = xay_1$ and $w_2 = xby_2$ for some $x, y_1, y_2 \in \Sigma^*$. Since $a \neq b$, we may assume that $a \in S_i$ but $b \notin S_i$ for some i, $1 \leq i \leq t$. Then there is a string $z \in \Sigma^*$ of length $im - |y_1| - 1$ such that $w_1 z \in L_n$ but $w_2 z \notin L_n$. Thus, $w_1 \not\equiv_{L_n} w_2$.

So, each word $w \in \Sigma^*$ such that $|w| \leq m$ represents a distinct equivalence class of $\equiv_{L_n}$. Hence, the total number of distinct equivalence classes of $\equiv_{L_n}$ is at least $(k^{m+1} - 1)/(k-1)$.

We could end the proof at this point. However, it is interesting to show also that we have counted all the distinct equivalence classes of $\equiv_{L_n}$, i.e., L_n is not a counterexample for Theorem 5. We give a proof only for the cases where $m = n/(t+1)$ is an integer.

We prove this by showing that, for each $w \in \Sigma^*$ such that $m < |w| < n$, there exists $u \in \Sigma^*$ with $|u| < |m|$ such that $w \equiv_{L_n} u$. Note that for each $w \in \Sigma^*$ such that $|w| \geq n$, we have $w \equiv_{L_n} a^m$, where $a \in \Sigma$ and $E_i(a) = 0$ for all i.

Let $w = a_l a_{l-1} \cdots a_2 a_1$. We need to consider the following two cases: (1) $tm < l < n$ and (2) $sm < l \leq (s+1)m$ for some s, $1 \leq s < t$.

For (1), we define the word $u = b_{tm} \cdots b_2 b_1$ such that $w \equiv_{L_n} u$ as follows: $b_i = a_i$ for all $1 \leq i \leq tm$ except $b_{l-m}, \ldots, b_{(t-1)m+1}$, where each b_j, $l - m \leq j \leq (t-1)m + 1$, is a symbol which has the same binary encoding as a_j with the possible exception that the tth bit of b_j's encoding is definitely 0. It can be shown that, for any $z \in \Sigma^*$ such that $|z| < n - l$, $wz \in L_n$ iff $uz \in L_n$; and, for any $z \in \Sigma^*$ such that $|z| \geq n - l$, $wz \notin L_n$ and $uz \notin L_n$.

For (2), we first define $u' = b_{sm} \cdots b_2 b_1$ as in (1) above. We then define $u = c_{sm} \cdots c_2 c_1$ such that $u_i = u'_i$ for all i, $1 \leq i \leq l$, but $i = l - m, \ldots, (s-1)m + 1$. Each c_j, $l - m \leq j \leq (s-1)m + 1$, has the same encoding as b_j except that, for $s \leq i < t$, the ith bit of c_j's encoding is 0 if the $(i+1)$th bit of b_{m+j}'s encoding is 0. It can be shown that $w \equiv_{L_n} u$.

Therefore, each word in Σ^* of length at most m represents a distinct equivalence class of $\equiv_{L_n}$ and there are no more distinct classes of $\equiv_{L_n}$. $\square$

In order to explain the above idea more clearly, we give the following example.

Example 1 Let $\Sigma = \{a, b, c, d\}$ and $n = 9$. Then $k = 4$, $t = \lceil \log_2 k \rceil = 2$, and $m = n/(t+1) = 3$. We encode the letters in Σ as follows:

$$a : 00, \quad b : 01, \quad c : 10, \quad d : 11$$

Then

$$S_1 = \{c, d\}, \quad S_2 = \{b, d\}$$

Define

$$L'_9 = \{x_2 a_2 x_1 a_1 x_0 \mid x_0, x_1, x_2 \in \Sigma^2, \ a_1 \in S_1, \ a_2 \in S_2\}.$$

Then

$$L_9 = \{w \in \Sigma^* \mid \exists x \in \Sigma^* \ xw \in L'_9 \ \& \ |w| \geq m\}.$$

It is easy to see that L_9 is accepted by an 9-state NFA. It can also be shown that each word $w \in \Sigma^*$ such that $|w| \leq 3$ represents a distinct equivalence class of $\equiv_{L_9}$ and there are no other distinct classes.

One can show that $adb \not\equiv_{L_9} acb$ because $adbcccc \in L_9$ and $acdcccc \notin L_9$. One can also show that

$$dddddddd \equiv_{L_9} dccddd \equiv_{L_9} caa$$

as well as

$$dcbdc \equiv_{L_9} bca.$$

When both $\log_2 k$ and $n/(\log_2 k + 1)$ are integers, the bound given by Theorem 5 can be reached and thus is optimal.

References

1. Hopcroft, J.E., Ullman, J.D.: *Introduction to Automata Theory, Languages, and Computation*. Addison Wesley, Reading, Mass., 1979.
2. Mandl, R.: Precise Bounds Associated with the Subset Construction on Various Classes of Nondeterministic Finite Automata. Proc. of the 7th Annual Princeton Conference on Information Science and Systems (1973) 263-267.
3. Meyer, A.R., Fischer, M.J.: Economy of description by automata, grammars, and formal systems. Proc. of FOCS 12 (1971) 188-191.
4. Myhill, J.: Finite automata and the representation of events. WADD TR-57-624, Wright Patterson AFB, Ohio, 1957, 112-137.
5. Nerode, A.: Linear automata transformation. Proceedings of AMS 9 (1958) 541-544.
6. Ravikumar, B., Ibarra, O.H.: Relating the type of ambiguity of finite automata to the succinctness of their representation. SIAM J. Comput. 18 (1989) 1263-1282.
7. Salomaa, A.: *Theory of Automata*. Pergamon Press, Oxford, 1969.
8. Salomaa, K., Yu, S., Zhuang, Q.: The state complexities of some basic operations on regular languages. Theoret. Comput. Sci. 125 (1994) 315-328.
9. Wood, D.: *Theory of Computation*. Wiley, New York, 1987.
10. Yu, S., Zhuang, Q.: On the state complexity of intersection of regular languages. ACM SIGACT News 22 no. 3 (1991) 52-54.

How to Use Sorting Procedures to Minimize DFA

Barbara Schubert

Department of Computer Science
The University of Western Ontario
London, Ontario, N6A 5B7, Canada

Abstract. In this paper we introduce a new idea, which can be used in minimization of a deterministic finite automaton. Namely, we associate names with states of an automaton and we sort them. We give a new algorithm, its correctness proof, and its proof of execution time bound. This algorithm has time complexity $O(n^2 \log n)$ and can be considered as a direct improvement of Wood's algorithm [6] which has time complexity $O(n^3)$, where n is the number of states. Wood's algorithm checks if pairs of states are distinguishable. It is improved by making better use of transitivity. Similarly some other algorithms which check if pairs of states are distinguishable can be improved using sorting procedures.

1 Introduction

The process of minimimizing Deterministic Finite Automaton (DFA) has many different applications, such as testing for faults in circuits [1]. The asymptotically fastest known algorithm for minimization is Hopcroft's algorithm [2], with a time complexity of $O(n \log n)$. Watson [5] has provided a complete list of all minimization algorithms.

Hopcroft's algorithm, while asymptotically fastest, is very complicated. In practice, less efficient, but less complex, algorithms are used, like Wood's algorithm [6] which has time complexity $O(n^3)$. We will show how to improve Wood's algorithm by using efficient sorting procedures. The upper bound for the version we will present is $O(n^2 \log n)$. We will explain how to apply the new version of the algorithm to any partition of the set of states (not only to the partition into final and nonfinal states), so that it can be used to directly solve the problems presented by Brzozowski and Jürgensen [1].

2. Basic Notions

As we want to define the algorithm for every possible partition of states, we will modify some definitions presented by Wood. For further details about this algorithm consult the text of Wood [6] and the monograph of Watson [5].

Definition 1. A *deterministic finite automaton (DFA)* M is specified by a quintuple $M = (Q, \Sigma, \delta, s, B)$ where

Q is the alphabet of *state* symbols;
Σ is the alphabet of *input* symbols;
$\delta : Q \times \Sigma \to Q$ is the *transition function*;
$s \in Q$ is the *start state*.
$B = (B_1, B_2 \ldots, B_r)$ is a partition of the states of the automaton.

As usual, we extend the transition function δ to a function of $Q \times \Sigma^*$ into Q by

$$\delta(q, w) = \begin{cases} q, & \text{if } w = \lambda, \\ \delta(\delta(q, x), w'), & \text{if } w = xw' \text{ with } x \in \Sigma, w' \in \Sigma^*, \end{cases}$$

for all $q \in Q$ and $w \in \Sigma^*$, where λ denotes the empty word. Instead of $\delta(q, w)$ we write qw. The partition of states of an automaton allows us to classify the input words. An input word $w \in \Sigma^*$ is of type B_i if $sw \in B_i$. Let ρ_B be the equivalence relation on Q defined by B. For $q \in Q$ we write $[q]_B$ to denote the class of q.

Definition 2. Let $M = (Q, \Sigma, \delta, s, B)$ be a DFA. Two states p and q of M are said to be *equivalent* if, for every $w \in \Sigma^*$, $[qw]_B = [pw]_B$. We write $p \equiv q$ in this case.

Two states p and q are *equivalent* if, given any input word w, executing the automaton on w leads to states belonging to the same block B_i.

Definition 3. Let $M = (Q, \Sigma, \delta, s, B)$, and $M' = (Q', \Sigma, \delta', s', B')$ be DFAs, and let $\mu : B \to B'$ be a bijection. M' μ-simulates M if, for every $q \in Q$, there is a $q' \in Q'$ such that, for every input word w in Σ^*, if $qw \in B_i$ then $q'w \in \mu(B_i)$. Also if $sw \in B_j$ then $s'w \in \mu(B_j)$. We write $M \leq_\mu M'$.

Definition 4. Let $M = (Q, \Sigma, \delta, s, B)$, and $M' = (Q', \Sigma, \delta', s', B')$ be DFAs. M and M' are *equivalent* if there is a bijection μ such that $M \leq_\mu M'$ and $M' \leq_{\mu^{-1}} M$. We write $M \simeq M'$

Definition 5. Let $M = (Q, \Sigma, \delta, s, B)$. M is *minimal* if, for all M' equivalent to M, $|Q| \leq |Q'|$.

The known algorithms for minimization of a DFA are usually explained using the partition of states into the two sets of final and nonfinal states, so $B = (F, Q \setminus F)$. For the partition into final and nonfinal states two automata are equivalent if and only if they accept the same language. The minimal automaton M' constructed from the automaton M should accept the same language as M and have the minimal number of states.

Thus one should be careful applying the above definitions and theorems to a specific example. It can be the case that only one mapping is permitted and so the definitions of the equivalence should be modified. For example given automata

$M = (Q, \Sigma, \delta, s, F)$ and $M' = (Q', \Sigma, \delta', s', F')$, where F and F' are the sets of final states, the only permitted map is μ where $\mu(F) = \mu(F')$ and $\mu(Q \setminus F) = Q' \setminus F'$. However, as far as minimization is concerned, this restriction is not relevant; if L is the language accepted by M then the minimal acceptors for L and $\Sigma^* \setminus L$ are isomorphic.

It can be noticed also that in the initial automaton usually the unreachable states are eliminated during the construction of the minimal automaton. The process of elimination of unreachable states is explained by Wood [6]. In the algorithm we will present, we will assume that all states are reachable.

As we want to apply the algorithm to every possible partition we should modify the definition of distinguishability and indistinguishability presented by Wood [6].

Definition 6. Let $M = (Q, \Sigma, \delta, s, B)$ be a DFA. For two distinct states $p, q \in Q$ we say p and q are *distinguishable* if there exists a $w \in \Sigma^*$ such that $[pw]_B \neq [qw]_B$. The word w *distinguishes* p from q. If no such word exists then p and q are *indistinguishable* or *equivalent*.

Given an integer $k \geq 0$, we say two distinct states $p, q \in Q$ are *k-distinguishable* if there is a word $w \in \Sigma^*$, $|w| \leq k$, which distinguishes p from q. If p and q are k-distinguishable we write $p \not\equiv^k q$. If there is no such word, then we say that p and q are *k-indistinguishable* and we write $p \equiv^k q$.

Clearly two states are distinguishable if they are k-distinguishable, for some $k \geq 0$. If we cannot distinguish p and q with words of length at most $k + 1$, for some $k \geq 0$, then we cannot distinguish them with words of length at most k. Thus, $p \equiv^{k+1} q$ implies $p \equiv^k q$, for all $k \geq 0$, that is, $\equiv^{k+1} \subseteq \equiv^k$. We say $\equiv^{k+1}$ is a *refinement* of $\equiv^k$. Similarly, we obtain $\equiv \subseteq \equiv^k$, for all $k \geq 0$.

3. When Sorting Is More Efficient

Wood's algorithm can be improved by making better use of transitivity. Given an automaton M, for some states p, q and z of this automaton, suppose that at step k of the algorithm we find out (p, q) and (q, z) are that the states in the pairs k-indistinguishable. By transitivity, we can conclude that the states (q, z) are also k-indistinguishable. So there is no need to check the pair (q, z) separately. One method that efficiently uses the notion of transitivity is sorting.

Sorting procedures can also distinguish states more efficiently. Assume that, at step k of the algorithm, unique names are associated with states so that sorting can be performed. Suppose that for some states p, q and z, the states in the pair (p, q) and the states in the pair (q, z) are k-distinguishable and as a result of sorting we obtain the list (p, q, z). This allows us to conclude that the states from the pair (p, z) are also k-distinguishable. This improvement applies not only to Wood's algorithm but also to Hopcroft and Ullman's algorithm [3].

4. A New Algorithm

Wood's algorithm for minimization works as follows: One starts with $\equiv^0 = \rho_B$, as two states are 0-indistinguishable if and only if they are in the same block of B. There are at most $r = |B|$ equivalence classes formed by $\equiv^0$. We construct $\equiv$ by refining $\equiv^0$ to give $\equiv^1$, $\equiv^1$ to give $\equiv^2$ and so on. This continues until $\equiv^k = \equiv^{k+1}$. Termination occurs no later than when $k = n - 2$, where $n = |Q|$.

The main idea used in the new version of this algorithm, is to give unique names to the equivalence classes formed at every step of the algorithm. At step k of the algorithm, every state p belonging to the same equivalence class will also have the same name, and states belonging to different classes will have different names. By N_p^k we denote the name of the equivalence class to which the state p belongs at step k of the algorithm. So at step k of the algorithm, if $p \equiv^k q$ then $N_p^k = N_q^k$, and if $p \not\equiv^k q$ then $N_p^k \neq N_q^k$. Whenever an equivalence class is changed (class is divided), the states in this class will get new names.

Suppose we are given a DFA $M = (Q, \Sigma, \delta, s, B)$. Let $\Sigma = (a_1, a_2, ..., a_m)$, where the input symbols are ordered in arbitrary but fixed way. With every state $p \in Q$, we associate the *pattern*

$$pattern(p^k) = \left(N_p^k, N_{pa_1}^k, N_{pa_2}^k, \; \ldots \; , N_{pa_m}^k \right)$$

at step k.

To avoid the calculation of patterns before every execution of sorting, states are represented by state structures: the state structure of a state p consists of the current name N_p^k of p and the transition list of p. The transition list of every state is implemented by pointers such that the pointer for pa_i in the state structure of p points to the state structure of the state pa_i. Now the algorithm can be presented.

Algorithm 7. *Input: A complete DFA M with set of reachable states Q, set of input symbols Σ, transitions function δ, start state s, and partition of states B.*

Output: A minimal DFA M' with state partition equivalent to M.
Method
1. *Associate the name N_{B_i} with every block B_i (so if $p \in B_i$ then $N_p^0 = N_{B_i}$)*
2. *Sort the names of states (construct $\equiv^k$ for $k = 0$)*
3. **repeat**
   > *sort patterns (construct $\equiv^{k+1}$)*
   > *if $\equiv^k \neq \equiv^{k+1}$ rename appropriate states*
   until $\equiv^k = \equiv^{k+1}$.

So, at every step $k + 1$ of the algorithm, new equivalence classes are formed and compared to the classes formed at step k. Renaming is done in the following way: Suppose that at step k of the algorithm, states p and q are in the same class and thus have the same name. If at step $k + 1$ states p and q have different patterns, they should be distinguished and so they should get different names at this step. If at step $k+1$ all states from some class have the same pattern, there is no need to change their names.

Example 8. We will illustrate the algorithm applied to the automaton in Fig.1. This is a minimal automaton and the example shows that the algorithm can verify this fact.

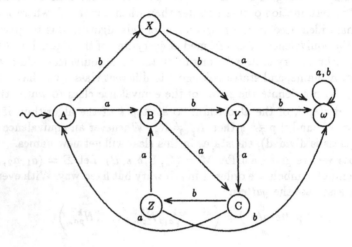

Fig. 1. Deterministic finite automaton M_1'

Suppose we are given the DFA M_1 with the partition of states of this automaton $B = (B_1, B_2, B_3) = (\{\omega\}, \{A, B, C\}, \{X, Y, Z\})$. Since B is a partition, each state p is in exactly one block B_j. We assign the initial name N_p^0 to each state $p \in Q$ as follows:
$$N_p^0 = j \text{ if } p \in B_j.$$
This gives the following initial names to every state:
$$N_\omega^0 = 1$$
$$N_A^0 = N_B^0 = N_C^0 = 2$$
$$N_X^0 = N_Y^0 = N_Z^0 = 3.$$

We will use a tabular method to compute $\equiv^{k+1}$ from $\equiv^k$ using sorting. We illustrate the state structure using the following format:

$$p^*(N_p^k) \quad pa^*(N_{pa}^k) \quad pb^*(N_{pb}^k)$$

where p^*, pa^*, and pb^* denote pointers (we will use the symbols from Q), and N_p^k, N_{pa}^k, and N_{pb}^k denote the current names of the states p, pa, and pb respectively. Remember that the numbers before the brackets (the pointers) and the brackets are not used for sorting purposes. Also notice that the comparisons are made between elements N_p^k which are included in the pattern, and not between other elements. So every N_p^k in the pattern is treated as one element, not as a sequence of symbols or letters. It should be remembered that individual N_p^k will have different lengths because of the renaming.

Table 1. The result of sorting at step 1 of the algorithm

$\equiv^0$	$p(N_p^1)$	$pa(N_{pa}^1)$	$pb(N_{pb}^1)$
class 1:	$\omega(1)$	$\omega(1)$	$\omega(1)$
class 2:	$A(2)$	$B(2)$	$X(3)$
	$C(2)$	$A(2)$	$Z(3)$
	$B(2)$	$C(2)$	$Y(3)$
class 3:	$X(3)$	$\omega(1)$	$Y(3)$
	$Y(3)$	$C(2)$	$\omega(1)$
	$Z(3)$	$B(2)$	$\omega(1)$

Table 2. The result of sorting at step 2 of the algorithm

$\equiv^1$		$p(N_p^1)$	$pa(N_{pa}^1)$	$pb(N_{pb}^1)$
class	1:	$\omega(1)$	$\omega(1)$	$\omega(1)$
class	2:	$A(2)$	$B(2)$	$X(31)$
		$B(2)$	$C(2)$	$Y(32)$
class 3.1:		$X(31)$	$\omega(1)$	$Y(32)$
class 3.2:		$Y(32)$	$C(2)$	$\omega(1)$
		$Z(32)$	$B(2)$	$\omega(1)$

Table 3. The result of sorting at step 3 of the algorithm

$\equiv^2$		$p(N_p^1)$	$pa(N_{pa}^1)$	$pb(N_{pb}^1)$
class	1:	$\omega(1)$	$\omega(1)$	$\omega(1)$
class 2.1:		$A(21)$	$B(22)$	$X(31)$
class 2.2:		$C(22)$	$A(21)$	$Z(32)$
		$B(22)$	$C(22)$	$Y(32)$
class 3.1:		$X(31)$	$\omega(1)$	$Y(32)$
class 3.2:		$Y(32)$	$C(22)$	$\omega(1)$
		$Z(32)$	$B(22)$	$\omega(1)$

Table 4. The result of sorting at step 4 of the algorithm

$\equiv^3$		$p(N_p^1)$	$pa(N_{pa}^1)$	$pb(N_{pb}^1)$
class	1:	$\omega(1)$	$\omega(1)$	$\omega(1)$
class	2.1::	$A(21)$	$B(22)$	$X(31)$
class 2.2.1:		$C(221)$	$A(21)$	$Z(32)$
class 2.2.2:		$B(222)$	$C(221)$	$Y(32)$
class	3.1:	$X(31)$	$\omega(1)$	$Y(32)$
class	3.2:	$Y(32)$	$C(221)$	$\omega(1)$
		$Z(32)$	$B(222)$	$\omega(1)$

The initial situation is shown in Table 1. After sorting the names of the states, we have only 3 equivalence classes. But when we sort the patterns, we see that $\{X\}$ and $\{Y, Z\}$ should be distinguished. So class 3 should be divided into two classes, and the names of the states should be changed to 3.1 and 3.2 (or 31 and 32) to get $\equiv^1$. We write a shorter horizontal line to denote that in the next step of the algorithm, the class should be divided, and so states in this class should be renamed. The results of the next sorting steps are shown in Table 2, Table 3, Table 4, and Table 5. Finally we see that $\equiv^4 = \equiv^5$ and so we cannot combine any states (no two states are equivalent) and the automaton M_1 is minimal.

Table 5. The result of sorting at step 5 of the algorithm

$\equiv^2$	$p(N_p^1)$	$pa(N_{pa}^1)$	$pb(N_{pb}^1)$
class 1:	$\omega(1)$	$\omega(1)$	$\omega(1)$
class 2.1:	$A(21)$	$B(22)$	$X(31)$
class 2.2.1:	$C(221)$	$A(21)$	$Z(322)$
class 2.2.2:	$B(222)$	$C(221)$	$Y(321)$
class 3.1:	$X(31)$	$\omega(1)$	$Y(321)$
class 3.2.1:	$Y(321)$	$C(221)$	$\omega(1)$
class 3.2.2:	$Z(322)$	$B(222)$	$\omega(1)$

Theorem 9. *Let $M = (Q, \Sigma, \delta, s, B)$ be a complete DFA having only reachable states and $M' = (Q', \Sigma, \delta', s', B')$ be the corresponding DFA constructed from $\equiv$ for M. Then M' is a minimal complete DFA equivalent to M.*

Proof. Assume M' is not minimal. Then there exists a minimal complete DFA $M'' = (Q'', \Sigma, \delta'', s'', B'')$ with $M \simeq M' \simeq M''$ and $|Q| \geq |Q'| > |Q''|$. Let $\mu : B' \to B''$ be a bijection which gives the equivalence of M' and M''. By associating each word $w \in \Sigma^*$ with the state it reaches from s'', we partition Σ^* into $|Q''|$ classes with respect to M''. The same holds for M' except that we obtain $|Q'|$ classes. Because M' only consists of reachable states, by construction, each class with respect to M' is nonempty. Because $|Q'| > |Q''|$ there must be words x and y such that $s''x = s''y$ and $s'x \neq s'y$. Let $p = s'x$ and $q = s'y$. So the states p and q must be distinguishable in M' (otherwise $p = q$ a contradiction). Hence, there is a z in Σ^* such that $[pz]_{B'} \neq [qz]_{B'}$. So $s'xz \in B_i'$, $s'yz \in B_j'$ for some $i \neq j$. As M' and M'' are equivalent $s''xz \in \mu(B_i')$ and $s''yz \in \mu(B_j')$. As $s''x = s''y$, therefore $s''xz = s''yz$ but as μ is a bijection $\mu(B_i') \neq \mu(B_j')$, which provides a contradiction, and establishes the theorem.

Theorem 10. *The runtime of Algorithm 7 is bounded from above by $n^2 \log n$.*

Proof. The algorithm terminates in at most $n - 2$ steps where $n = |Q|$. At every step one sorting procedure is executed, which requires at most $n \log n$ comparisons. This gives the upper bound of $O(n^2 \log n)$.

It is difficult to describe precisely which automaton will represent the worst case, as the more comparisons we make at every step of the algorithm, the fewer potential sorting procedures we will need to call in the future. Also, during consecutive steps of sorting, the list is more "in order", so fewer comparisons are used to update it.

This algorithm is based on the sorting of states, so the minimal time for the algorithm can't be less than the time required to sort the list of patterns once (to check if $\equiv^0 = \equiv^1$). It will be $O(n \log n)$ if and only if all patterns are distinct. Otherwise it depends on the number of distinct patterns. Consider an automaton which has only one final state and $n - 1$ nonfinal states, where all nonfinal states are combined into one state in the minimal automaton. This corresponds to having n elements to sort, where only one is different from all others. This implies that patterns are sorted just once. Using mergesort when we merge any 2 lists, we can combine the "same" elements so the maximum number of elements in the lists to be merged is 2 and the maximum number of comparisons is $n + \log n$. This implies that the best case for this algorithm has time complexity $O(n)$.

It is possible to improve Algorithm 7 in such a way that it will have time complexity $O(n^2)$. We will achieve this time complexity by requiring that the list of patterns will be sorted class by class. The additional requirement, that only the classes which will be divided at the next step are sorted, gives the same time complexity. The details about these new versions of the algorithm will be included in Schubert's MSc Thesis [4].

References

1. J. A. Brzozowski, H. Jürgensen: A model for Sequential Machine Testing and Diagnosis, *Journal of Electronic Testing: Theory and Applications* **3** (1992), 219–234.
2. J. E. Hopcroft: An $n \log n$ algorithm for minimizing the states in a finite automaton, in: Z. Kohavi, ed., *The Theory of Machines and Computations*, Academic Press, New York, 1971, 189–196.
3. J. E. Hopcroft, J. D. Ullman: *Introduction to Automata Theory, Languages and Computation*, Addison-Wesley, New Jersey , March 1979.
4. B. Schubert: *How to Use Sorting Procedures to Minimize DFA*, MSc Thesis, Department of Computer Science, University of Western Ontario, in preparation.
5. B. W. Watson: *Taxonomies and Toolkits of Regular Language Algorithms*, Eindhoven University of Technology, The Netherlands, 1995.
6. D. Wood: *Theory of Computation*, John Wiley & Sons, New York, 1987.

FIRE Lite: FAs and REs in C++

Bruce W. Watson

Ribbit Software Systems Inc., IST Technologies Research Group
Box 24040, 297 Bernard Ave., Kelowna, B.C., V1Y 9P9, Canada
e-mail: watson@RibbitSoft.com, fax: +1 514 938 9308

Abstract. This paper describes a C++ finite automata toolkit known
as FIRE Lite (FInite automata and Regular Expressions; Lite since it is
the smaller and newer cousin of the FIRE Engine toolkit, also originally
developed at the Eindhoven University of Technology). The client interface and aspects of the design and implementation are also described.
The toolkit includes implementations of almost all of the known automata construction algorithms and many of the deterministic automata
minimization algorithms. These implementations enabled us to collect
performance data on these algorithms. The performance data, which we
believe to be the first extensive benchmarks of the algorithms, are also
summarized in this paper.

1 Introduction and related work

FIRE Lite is a C++ toolkit implementing finite automata and regular expression
algorithms. The toolkit is a computing engine, providing classes and algorithms
of a low enough level that they can be used in most applications requiring finite
automata or regular expressions. Almost all of the algorithms derived in [Wat95,
Chapter 6] are implemented. We begin by considering other toolkits.

1.1 Related toolkits

There are several existing finite automata toolkits. Some of the more closely
related ones are described here:

- The FIRE Engine II system, as described at the Ribbit Software Systems
  Inc. web site, www.RibbitSoft.com. The FIRE Engine II is a commercial re-
  implementation of FIRE Lite. It includes a number of features not found in
  the FIRE Lite, such as: regular grammars (in addition to regular expressions),
  extended regular expressions, Mealy transducers (including their minimization), template parameterized transition labels of input and output alphabets, predicate transitions, dictionaries, and a full class library with multiple
  implementations providing different trade-offs with respect to running time
  and space. The performance of the implementation makes the toolkit suitable for use in natural language processing (involving automata with millions
  of states) and in compilers, to name a few applications. The FIRE Engine II
  is, however, much larger than FIRE Lite, making FIRE Lite more suitable for
  study by students of algorithmics, software engineering and programming.

- The FIRE Engine, as described in [Wat94a, Wat94b]. The FIRE Engine was the first of the toolkits from the Computing Science Faculty in Eindhoven. It is an implementation of all of the algorithms appearing in two early taxonomies of finite automata algorithms which appeared in [Wat93a, Wat93b]. The toolkit is somewhat larger than FIRE Lite (the FIRE Engine is 9000 lines of C++) and has a slightly larger and more complex public interface. The more complex interface means that the toolkit does not support multi-threaded use of a single finite automaton.
- The Grail system, as described in [RW93]. Grail follows in the tradition of such toolkits as Regpack [Leis77] and INR [John86], which were all developed at the University of Waterloo, Canada. It provides two interfaces:
  - A set of 'filter' programs (in the tradition of UNIX). Each filter implements an elementary operation on finite automata or regular expressions. Such operations include conversions from regular expressions to finite automata, minimization of finite automata, etc. The filters can be combined as a UNIX 'pipe' to create more complex operations; the use of pipes allows the user to examine the intermediate results of complex operations. This interface satisfies the first two (of three) aims of Grail [RW93]: to provide a vehicle for research into language theoretic algorithms, and to facilitate teaching of language theory.
  - A raw C++ class library provides a wide variety of language theoretic objects and algorithms for manipulating them. The class library is used directly in the implementation of the filter programs. This interface is intended to satisfy the third aim of Grail: an efficient system for use in application software.
- The Amore system, as described in [JPTW90]. The Amore package is an implementation of the semigroup approach to formal languages. It provides procedures for the manipulation of regular expressions, finite automata, and finite semigroups. The system supports a graphical user-interface on a variety of platforms, allowing the user to interactively and graphically manipulate the finite automata. The program is written (portably) in the C programming language, but does not provide a programmer's interface. The system is intended to serve two purposes: to support research into language theory, and to help explore the efficient implementation of algorithms solving language theoretic problems.
- The Automate system, as described in [CH91]. Automate is a package for the symbolic computation on finite automata, extended regular expressions (those with the intersection and complementation operators), and finite semigroups. The system provides a textual user-interface through which regular expressions and finite automata can be manipulated. A single finite automata construction algorithm (a variant of Thompson's) and a single deterministic finite automata minimization algorithm is provided (Hopcroft's). The system is intended for use in teaching and language theory research. The (monolithic) program is written (portably) in the C programming language, but provides no function library interface for programmers.

According to Pascal Caron (at the Université de Rouen, France), a new version of Automate is being written in the MAPLE symbolic computation system.

The provision of the C++ class interface in Grail makes it the only toolkit with aims similar to those of the FIRE Engine and of FIRE Lite. In the following section, we will highlight some of the advantages of FIRE Lite over the other toolkits.

1.2 Advantages and characteristics of FIRE Lite

The advantages to using FIRE Lite, and the similarities and differences between FIRE Lite and the existing toolkits are:

- FIRE Lite does not provide a user interface[1]. Some of the other toolkits provide user interfaces for the symbolic manipulation of finite automata and regular expressions. Since FIRE Lite is strictly a computing engine, it can be used as the implementation beneath a symbolic computation application.
- The toolkit is implemented for efficiency. Unlike the other toolkits, which are implemented with educational aims, it is intended that the implementations in FIRE Lite are efficient enough that they can be used in production quality software. This means that we have chosen to use compile-time constructs such as templates instead of inheritance and virtual methods. The use of templates and their methods can be optimized by the C++ compiler, whereas virtual function calls (and inheritance hierarchies) are inherently run-time constructs.
- Despite the emphasis on efficiency in FIRE Lite the toolkit still has educational value. The toolkit bridges the gap between the easily understood abstract algorithms appearing in [Wat95, Chapter 6] and practical implementation of such algorithms. The C++ implementations of the algorithms display a close resemblance to their abstract counterparts.
- Most of the toolkits implement only one of the known algorithms for constructing finite automata. For example, Automate implements only one of the known constructions. By contrast, FIRE Lite provides implementations of almost all of the known algorithms for constructing finite automata. (See [Wat95, Chapter 6] for a taxonomy of the algorithms.) Implementing many of the known algorithms has several advantages:
  - The client can choose between a variety of algorithms, given tradeoffs for finite automata construction time and input string processing time.
  - The efficiency of the algorithms (on a given application) can be compared.
  - The algorithms can be studied and compared by those interested in the inner workings of the algorithms.

---

[1] A rudimentary user interface is included for demonstration purposes.

2   Using the toolkit

In this section, we describe the client interface to the toolkit — including some examples which use the toolkit. The issues in the design of the current client interface are detailed in [Wat95].

There are two components to the client interface of FIRE Lite: regular expressions (class *RE*) and finite automata (classes whose names begin with *FA...*). We first consider regular expressions and their construction.

Regular expressions are implemented through class *RE*. This class provides a variety of constructors and member functions for constructing complex regular expressions. Stream insertion and extraction operators are also provided. (The public interface of *RE* is rather fat, consisting of a number of member functions intended for use by the constructors of finite automata.) The following program constructs a simple regular expression and prints it:

```
#include "re.hpp"
#include <iostream.h>

int main(void) {
    auto RE e( 'B' );
    auto RE f( CharRange( 'a', 'z' ) );
    e.concatenate( f );
    e.or( f );
    e.star();
    cout << e;
    return( 0 );
}
```

The header `re.hpp` defines regular expression class *RE*. The program first constructs two regular expressions. The first is the single symbol *B*. The second regular expression is a *CharRange* — a character range. The *RE* constructed corresponds to the range $[a, z]$ of characters. No particular ordering is assumed on the characters, though most platforms use the ASCII ordering. Character ranges can always be used as atomic regular expressions. The program concatenates regular expression *f* onto *e* and then unions *f* onto *e*. Finally, the Kleene closure operator is applied to regular expression *e*. The final regular expression, which is $((B \cdot [a, z]) \cup [a, z])^*$, is output (in a prefix notation) to standard output.

The abstract finite automata class defines the common interface for all finite automata; it is defined in `faabs.hpp`. A variety of concrete finite automata are provided in FIRE Lite; they are declared in header `fas.hpp`. There are two ways to use a finite automaton. In both of them the client constructs a finite automaton, using a regular expression as argument to the constructor. The two are outlined as follows:

1. In the simplest of the two, the client program calls finite automaton member function *FAAbs::attemptAccept*, passing it a string and a reference to an integer. The member function returns *TRUE* if the string was accepted by

the automaton, *FALSE* otherwise. Into the integer reference it places the index (into the string) of the symbol to the right of the last symbol processed.

2. In the more complex method, the client takes the following steps (which resemble the steps required in using a pattern matcher mentioned in [Wat95, Chapter 9]):

   (a) The client calls the finite automaton member function *FAAbs::reportAll*, passing it a string and a pointer to a function which takes an integer and returns an integer. As in the SPARE Parts (the keyword pattern matching toolkit which is a companion to FIRE Lite — see [Wat95, Chapter 9]), the function is the 'call-back' function.

   (b) The member function processes the input string. Each time the finite automaton enters a final state (while processing the string), the call-back function is called. The argument to the call is the index (into the input string) of the symbol immediately to the right of the symbol which took the automaton to the final state.

   (c) To continue processing the string, the call-back function returns *TRUE*, otherwise *FALSE*.

   (d) When the input string is exhausted, or the call-back function returns *FALSE*, or the automaton becomes stuck (unable to make a transition on the next input symbol), the member function *FAAbs::reportAll* returns the index of the symbol immediately to the right of the last symbol on which a successful transition was made.

The following program fragment takes a regular expression, constructs a finite automaton, and processes an input string:

```
#include "com-misc.hpp"
#include "re.hpp"
#include "fas.hpp"
#include <iostream.h>

static int report( int ind ) {
    cout << ind << '\n';
    return( TRUE );
}

void process( const RE& e ) {
    auto FARFA M( e );
    cout << M.reportAll( "hishershey", &report );
    return;
}
```

Header `com-misc.hpp` provides the definition of constants *TRUE* and *FALSE* (these will be eliminated with the new **bool** C++ datatype), while header `fas.hpp` gives the declarations of a number of concrete finite automata. Function *report* is used as the call-back function; it simply prints the index and returns *TRUE* to continue processing. Function **process** takes an *RE* and constructs a local finite automaton (of concrete class *FARFA*). It then uses the automaton processes string **hishershey**, writing the final index to standard output before returning.

Given these examples, we can now demonstrate a more complex use of a finite automaton. In this example, we implement a generalized Aho-Corasick pattern matcher (GAC — as in [Wat95, Chapter 5]) which performs regular expression pattern matching. Since regular expression pattern matching is not presently included in the SPARE Parts, this example illustrates how FIRE Lite could be used to implement such pattern matching for a future version of the SPARE Parts. (This example also highlights the fact that the call-back mechanism in FIRE Lite is very similar to the mechanism in the SPARE Parts.)

```
#include "re.hpp"
#include "fas.hpp"
#include "string.hpp"

class PMRE {
public:
    PMRE( const RE& e ) : M( e ) {}
    int match( const String& S, int cb( int ) ) {
        return( M.reportAll( S, cb ) );
    }
private:
    FARFA M;
};
```

10

Headers **re.hpp**, **fas.hpp**, and **string.hpp** have all been explained before. Class *PMRE* is the regular expression pattern matching class. Its client interface is modeled on the pattern matching client interfaces used in [Wat95, Chapter 9]. The class has a private finite automaton (in this case an *FARFA*) which is constructed from an *RE*. (Note that the constructor of class *PMRE* has an empty body, since the constructor of *FARFA M* does all of the work.) The member function *PMRE::match* takes an input string and a call-back function. It functions in the same way as the call-back mechanism in [Wat95, Chapter 9]. The member function is trivial to implement using member *FAAbs::reportAll*. Whenever the finite automaton enters an accepting state, a match has been found and it is reported.

An important feature of FIRE Lite (like SPARE Parts) is that the call-back client interface implicitly supports multi-threading. See [Wat95, Chapter 9] for a discussion of call-back functions and multi-threading.

3 The structure of FIRE Lite

In this section, we give an overview of the structure of FIRE Lite and some of the main classes in the toolkit.

In the construction algorithms of [Wat95, Chapter 6], the finite automata that are produced have states containing extra information. In particular, the 'canonical construction' produces automata whose states are dotted regular expressions, or items. Some of the other constructions produce automata with sets of items for states, or sets of positions for states, or derivatives for states, and so on.

The constructions of [Wat95, Chapter 6] appear as the constructors (taking a regular expression) of various concrete finite automata classes in FIRE Lite. It seems natural that mathematical concepts such as items, sets of items, positions, and sets of positions will also appear in such an implementation. The only potential problem is the performance overhead inherent in implementing automata with states which are sets of items, etc.

The solution used in FIRE Lite is to implement states with internal structure as *abstract-states* during the construction of a finite automaton. The finite automaton is constructed using the abstract-states (so that the constructor corresponds to one of the algorithms in [Wat95, Chapter 6]). Once the automaton is fully constructed, the abstract-states are too space and time consuming for use while processing a string (for acceptance by the automaton). Instead, we map the abstract-states (and the transition relations, etc) to *States* (which are simply integers) using a *StateAssoc* object. Once the mapping is complete, the abstract-states, their transition relations, and the *StateAssoc* object can be destroyed.

For all of the finite automata which are constructed from abstract-states, the constructor takes an initial abstract-state, which is used as a 'seed' for constructing the remaining states and the transition relation. For performance reasons, we wish to make the finite automaton constructor a template class whose template argument is the abstract-state class (this would avoid virtual function calls). Unfortunately, most compilers which are presently available do not support template member functions (which have recently been added to the draft C++ standard). As a result, we are forced to make the entire finite automata class into a template class. The main disadvantage is that this reduces the amount of object code sharing (see [Wat95, Chapter 8] for a discussion of the differences between source and object code sharing and templates versus inheritance).

Most of the abstract-state classes have constructors which take a regular expression. As a result, an *RE* can be used as argument to most of the finite automata classes — a temporary abstract-state will be constructed automatically.

There are three types of abstract-states, corresponding to finite automata with ε-transitions (see class *FAFA*), ε-free finite automata, and deterministic finite automata. Classes of a particular variety of abstract-state all share the same public interface by convention; the template instantiation phase of the compiler detects deviations from the common interface. For more on the three types of abstract-state, see [Wat95, Chapter 10].

The transition relations on *States* are implemented by classes *StateStateRel* (for ε-transitions), *TransRel* (for nondeterministic transitions), and *DTransRel* (for deterministic transitions).

The following table presents a summary of the various concrete automata classes and their template arguments (if any):

Class	Description	
FACanonical	Canonical automaton	
FAFA	Automaton template, with ε-transitions	
	Single template argument required:	
	Abstract states	
	ASItems	Item sets (canonical)
FARFA	ε-*free* automaton	
FAEFFA	Automaton template, without ε-transitions	
	Single template argument required:	
	Abstract states	
	ASEFItems	Items sets (no filter)
	ASEFPosnsBS	Berry-Sethi
	ASEFPosnsBSdual	dual of Berry-Sethi
	ASEFPDerivative	Antimirov
FADFA	Deterministic automaton template	
	Single template argument required:	
	Abstract states	
	ASDItems	Items sets (no filter)
	ASDItemsDeRemer	Items sets (DeRemer filter)
	ASDItemsWatson	Items sets (W filter)
	ASDPosnsMYG	McNaughton-Yamada-Glushkov
	ASDPosnsASU	Aho-Sethi-Ullman
	ASDDerivative	Brzozowski

4 The performance of FIRE Lite

In this section, we present performance data for the algorithms implemented in FIRE Lite. We begin by describing the testing methodology, followed by the performance of automata construction algorithms, and the time that each automaton type requires for a single transition; we conclude with the performance of the minimization algorithms.

4.1 Testing methodology

This section gives an overview of the methods used in gathering the test data. We begin with the details of the test environment, followed by the details of the methods used to generate the regular expressions for input to the algorithms.

Test environment All of the tests were performed on an IBM-compatible personal computer running MS-Dos. The machine has an INTEL PENTIUM processor with a 75 Mhz clock, an off-chip cache of 256 kilobytes and main memory of 8 megabytes. During all of the tests, no other programs which could consume processing power were installed.

The test programs were compiled with the WATCOM C++32 compiler (version 9.5a) with optimizations for speed. The WATCOM compiler is bundled with an MS-DOS extender (used to provide virtual memory for applications with large data-structures) known as DOS/4GW. Since the use of virtual memory could affect the performance data, all data-structures were made to fit in physical memory.

Timing was done by reprogramming the computer's built-in counter to count microseconds. This reprogramming was encapsulated within a C++ timer class which provided functionality such as starting and stopping the timer. Member functions of the class also subtracted the overhead of the reprogramming from any particular timing run.

Generating regular expressions A large number of regular expressions were randomly generated, from which the finite automata were built. We used the random number generator appearing in [PTVF92, p. 280]. The regular expressions were generated as follows:

1. A height in the range [2, 5] was randomly chosen for the parse tree of the regular expression. (Larger heights were not chosen for memory reasons; smaller heights were not chosen since the constructions were performing close to the clock resolution.)
2. A regular expression of the desired height was generated, choosing between all of the eligible operators[2] with equal probability.
3. For the leaves, Ø and ε nodes were never chosen. The Ø was omitted, since such regular expressions prove to be uninteresting (they simply denote the empty language). Similarly, the ε was omitted, since the same effect is obtained by generating ? nodes[3].

There are a total of 3462 REs; the mean size is 9.63 nodes and the standard deviation is 4.72. The distribution of the number of nodes reflects the way in which the regular expressions were generated. Note that there are no regular expressions with a single node or with three nodes. These were omitted since the time to construct an FA from such small REs was usually below the resolution of the timer. (Two node regular expressions were used since they contain a * or a + node at the root. All of the constructions require more time to construct an automaton corresponding to such an expression.) Furthermore, it is not possible to generate lengthy strings in the language of such regular expressions.

Other statistics on the regular expressions were also collected, such as the height of the REs, the star-height of the REs, and some measure of the inher-

---

[2] Some operators will not be eligible. For example, to generate an RE of height 3, only the unary or binary operators can appear at the root.

[3] The ? operator is the regular operator used to indicate when something is optional.

ent nondeterminism in the REs[4]. These statistics did not prove to be useful in considering the performance of the algorithms.

Generating input strings For each type of automaton (FA, ε-free FA, and DFA), we also present data on the time required to make a single transition. In order to measure this, for each RE used as input to the constructions we generate a string in the prefix of the language denoted by the RE. Strings of length up to 10,000 symbols were generated.

The constructed automaton processes the string, making transitions, while the timer is used to measure the elapsed time. The time was divided by the number of symbols processed, yielding the average time for a single transition.

4.2 FA construction algorithm performance

In this section, we consider the time required (by each of the construction algorithms) to construct an automaton from a regular expression.

The algorithms The algorithms tested were derived in [Wat95, Chapter 6]. They have also been implemented in FIRE Lite. The implementations are discussed in detail in [Wat95]. For convenience, we will put the algorithms in two groups: those producing an FA, and those producing a DFA. The FA constructions are:

- The canonical construction (TH, since it is a variant of Thompson's construction), appearing in [Thom68] and [Wat95, Construction 6.15].
- The Berry-Sethi construction (BS), given in [BS86] and [Wat95, Construction 6.39].
- The dual of the Berry-Sethi construction (BS-D), appearing as [Wat95, Construction 6.65].

The DFA constructions are:

- The Aho-Sethi-Ullman construction (ASU) — [ASU86] and [Wat95, Construction 6.86].
- The deterministic item set construction (IC) — [Wat95, Example 6.23].
- DeRemer's construction (DER) — [Wat95, page 159].
- The filtered item set construction (FIC) — [Wat95, page 158].
- The McNaughton-Yamada-Glushkov construction (MYG), given as [MY60] and [Wat95, Construction 6.44].

For specific information on these algorithms, see [Wat95, Chapter 6].

Noticeably absent from this list are the derivatives-based algorithms (Brzozowski's and Antimirov's algorithms). The derivatives in these algorithms are

---

[4] One estimate of such nondeterminism is the ratio of alternation (union) nodes and * or + nodes to the total number of nodes in the regular expression.

used as the states. In their pure forms, the derivatives are stored as regular expressions. The space and time required to store and manipulate the regular expressions proved to be extremely costly, when compared to the representations of states used in some of the other constructions. The derivative-based algorithms consistently performed 5 to 10 times slower than the next slowest algorithm (the IC algorithm, in particular). Indeed, the preliminary testing could only be done for the smallest regular expressions without making use of virtual memory (which would further degrade their performance). No doubt the use of clever coding tricks would improve these algorithms greatly — though such coding tricks would yield a new algorithm.

Construction times For each of the generated regular expressions and each of the constructions, we measured the number of microseconds required to construct the automaton. For many applications, the construction time can be the single biggest factor in determining which construction to use.

First, the constructions are divided into two groups: those producing an FA (not a DFA), and those producing a DFA. The median performance of these two groups of constructions was graphed against the number of nodes in the regular expressions in Figures 1 and 2 respectively. All three of the algorithms

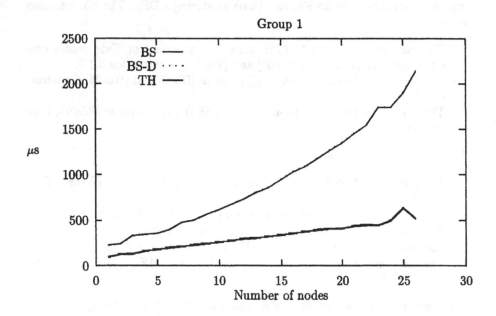

Fig. 1. Median construction times for FA constructions graphed against the number of nodes in the regular expressions. Note that BS and BS-D are superimposed as the higher ascending line.

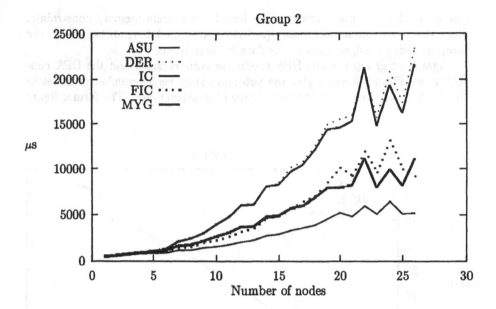

Fig. 2. Median construction times for DFA constructions graphed against the number of nodes in the regular expressions. The lowest line is ASU performance, while the middle pair of lines are MYG and FIC; the highest pair of lines is DER and IC.

in the first group are predicted to perform linearly in the number of nodes in the input regular expression. In Figure 1, the performance of the BS and BS-D constructions was nearly identical. They both performed somewhat worse than the TH algorithm. The apparent jump in construction time (of the TH algorithm) for 25 node regular expressions is due to the fact that only a single such expression was generated. Had more regular expressions been generated, the median performance would have appeared as a straight line (following the linear performance predicted for the TH algorithm).

The scale on Figure 2 shows that the second group of constructions was much slower than the first group. The ASU construction was by far the fastest, with FIC and MYG being the middle performers, and DER and IC being the slowest. In the range of 20 to 26 nodes, all of the constructions displayed peaks in their construction times. In these cases, some of the generated regular expressions have corresponding DFAs which are exponentially larger — forcing all of the constructions to take longer.

Constructed automaton sizes The size of each of the constructed automata was measured. The amount of memory space consumed by an automaton is directly proportional to the number of states in the automaton, and this data can be used to choose a construction based upon some memory constraints. Since the exact amount of memory (in bytes) consumed depends heavily on the compiler being used, we present the data in state terms.

Again, we group the non-DFA producing constructions and the DFA constructions. Figures 3 and 4 give the automata sizes versus number of nodes in the regular expressions for the two groups of constructions. The former figure

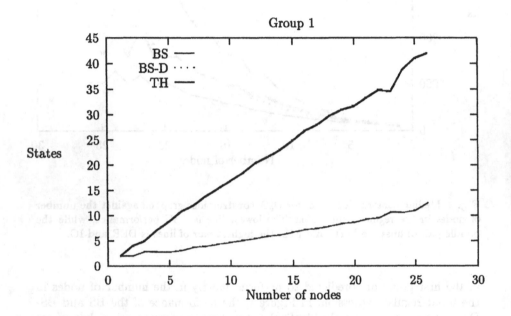

Fig. 3. The number of states in the FA is graphed against the number of nodes in the regular expression used as input for the TH, BS, and BS-D constructions. Note that the BS and BS-D constructions produce automata of identical size.

shows that the size of the TH-generated automata can grow quite rapidly. The BS and BS-D constructions produce automata of identical size In the second figure (Figure 4), we can identify two interesting properties of the constructions: ASU and FIC produce automata of the same size, as do the pair IC and MYG. Given the superior performance of ASU (over FIC), there is little reason to make use of FIC. Similarly, the MYG construction out-performs IC, and there is no reason to use IC since the automata will be the same size.

4.3 Automaton transition performance

The time required to make a single transition was measured for FAs (using TH), ε-*free* FAs (using BS), and DFAs (using FIC). The median time for the

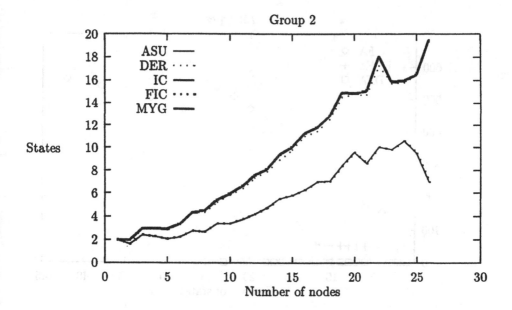

Fig. 4. The number of states in the DFA is graphed against the number of nodes in the regular expression used as input for the ASU, DER, FIC, IC, and MYG constructions. Note that the FIC and ASU constructions produce DFAs of identical size (the superimposed lower line), as did the pair IC and MYG (the superimposed higher line).

transitions has been graphed against the number of states in Figure 5. (Note that, for a given number of states, the number of each of the different types of automata varied.) The more general FAs displayed the slowest transition times, since the current set of states is stored as a set of integers, and the ε-transition and symbol transition relations are stored in a general manner. The ε-free FAs displayed much better performance, largely due to the time required to compute ε-transition closure in an automaton with ε-transitions. With the simple array lookup mechanism used in DFAs, it is not surprising that their transitions are by far the fastest (below one microsecond) and are largely independent of the number of states in the DFA.

The variance for general FAs and ε-*free* FAs is quite large. In both cases, the time for a single transition depends upon the number of states in the current state set. The variance for a DFA transition was also quite large. This is largely due to the fact that single transitions are around the resolution of the timer, meaning that other factors play a role. Such factors include artificial ones such as clock jitter and real ones such as instruction cache misses.

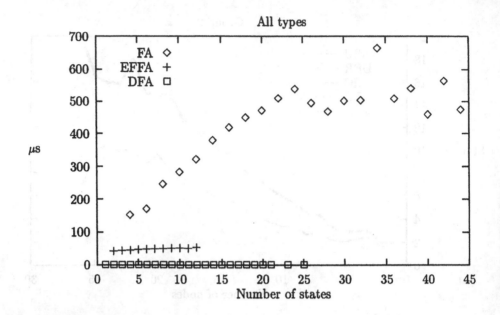

Fig. 5. Median time to make a single transition in microseconds (μs), for FAs, ε-free FAs, and DFAs, versus the number of states.

4.4 Automata construction algorithm recommendations

The conclusions relating to automata construction algorithms are:

- The advantages of the 'filters' introduced in [Wat95, Chapter 6] can be seen in the performance of IC, which is worse than either DER or FIC. Furthermore, IC produces larger automata than either DER or FIC.
- Predictably, the subset construction (with start-unreachable state removal) is a costly operation. All of the DFA construction algorithms were significantly slower than the general FA constructions.
- ASU is the subset construction composed with BS-D, while MYG is the subset construction composed with BS. Given the identical performance of BS and BS-D, it is interesting to note that ASU was significantly faster than MYG. Furthermore, ASU produced smaller DFAs than MYG.
- The time to make a transition can vary widely for different types of automata. As such, it can be an important factor in choosing a type of automaton.
    - Both FAs with ε-transitions and ε-free FAs have transition times that depend upon the number of states in the automaton; however, FAs with ε-transitions have significantly longer transition times.

- DFAs have transition times which are largely independent of the size of the automaton[5]. These times were around the resolution of the clock[6].

4.5 DFA minimization algorithms

Very little is known about the performance of DFA minimization algorithms in practice. Most software engineers choose an algorithm by reading their favourite text-book, or by attempting to understand Hopcroft's algorithm — the best known algorithm, with $O(n \log n)$ running time. The data in this section will show that somewhat more is involved in choosing the right minimization algorithm. In particular, the algorithms appearing in a popular formal languages text-book [HU79] and in a compiler text-book [ASU86] have relatively poor performance (for the chosen input data). Two of the algorithms which could be expected to have poor performance[7] actually gave impressive results in practice. Recommendations for software engineers will be also be given.

The algorithms The five algorithms are fully derived and taxonomized in [Wat95, Chapter 7] and their corresponding implementations are described in detail in [Wat95, Chapter 11]. Here, we give a brief summary of each of the algorithms (and an acronym which will be used in these sections to refer to the algorithm):

- The Brzozowski algorithm (BRZ), first derived in [Brzo62]. It is unique in being able to process a FA (not necessarily a DFA), yielding a minimal DFA which accepts the same language as the original FA.
- The Aho-Sethi-Ullman algorithm (ASU), appearing in [ASU86]. It computes an equivalence relation on states which indicate which states are indistinguishable. The appearance of this algorithm in [ASU86] has made it one of the most popular algorithms among implementors.
- The Hopcroft-Ullman algorithm (HU), appearing in [HU79]. It computes the complement of the relation computed by the Aho-Sethi-Ullman algorithm. Since this algorithm traverses transitions in the DFA in the reverse direction, it is at a speed disadvantage in most DFA implementations (including the one used in FIRE Lite and the FIRE Engine).
- The Hopcroft algorithm (HOP), presented in [Hopc71]. This is the best known algorithm (in terms of running time analysis) with running time of $O(n \log n)$ (where n is the number of states in the DFA).

---

[5] Although inspecting the implementation reveals that very dense transition graphs will yield more costly transitions than a sparse transition graph.

[6] This does not invalidate the results, since the average transition time is taken over a large number of transitions.

[7] They are Brzozowski's algorithm (which uses the costly subset construction) and a new algorithm which has exponential worst-case running time.

- The new algorithm (BW) appearing as [Wat95, Algorithm 7.28]. It computes the equivalence relation (on states) from below (with respect to the refinement ordering). The practical importance of this is explained in [Wat95, Chapter 7].

These algorithms are presently the only ones implemented in FIRE Lite.

Results For each of the random DFAs and each of the five algorithms, we measured the time in microseconds (μs) required to construct the equivalent (accepting the same language) minimal DFA.

The performance of the algorithms was graphed against the number of states in the original DFA. Graphing the performance against the number of edges in the original DFA was not found to be useful in evaluating the performance of the algorithms. The algorithms can be placed in two groups, based upon their performance. In order to aid in the comparison of the algorithms, we present graphs for these two groups separately.

The first group (ASU and HU) are the slowest algorithms; the graph appears in Figure 6. The HU algorithm was the worst performer of the five algorithms. It traverses the transitions (in the input DFA) in the reverse direction. A typical implementation of a DFA does not favour this direction of traversal. The ASU algorithm performed slightly better, although its performance is also far slower than any of BRZ, HOP, or BW. The second group (BRZ, HOP, and BW) is significantly faster; the corresponding graph appears in Figure 7. Note that this graph uses a different scale from the one in Figure 6. The data point (for 17 states) for the BW algorithm was dropped, since (at that point) the algorithm was more than 30 times slower than any of the other algorithms in this group (this is in keeping with the exponential worst-case running time of the algorithm).

4.6 Minimization algorithm conclusions and recommendations

We can draw the following conclusions from the data presented in these sections:

- Given their relative performance, the five algorithms can be put into two groups: the first consisting of ASU and HU, and the second consisting of BRZ, HOP, and BW.
- The HU algorithm has the lowest performance of all of the algorithms. This is largely due to the fact that it traverses the transitions of the DFA in the reverse direction — a direction not favoured by most practical implementations of DFAs.
- The ASU algorithm also displays rather poor performance. Traditionally, the algorithm has been of interest because it is easy to understand. The simplicity of BRZ minimization algorithm makes it even more suitable for teaching purposes.
- The HOP algorithm is the best known algorithm (in term of theoretical running time), with $\mathcal{O}(n \log n)$ running time. Despite this, it is the worst of

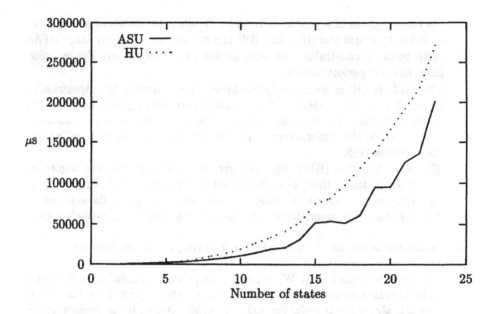

Fig. 6. Median performance (in microseconds to minimize) versus DFA size.

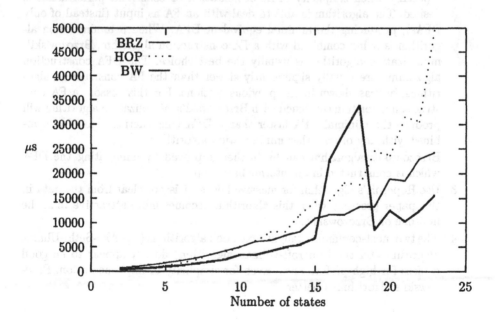

Fig. 7. Median performance (in microseconds to minimize) versus DFA size.

the second group of algorithms. With its excellent theoretical running time, it will outperform the BRZ and BW algorithms on extremely large DFAs. With memory constraints, we were unable to identify where the crossing-point of their performance is.

- The BRZ algorithm is extremely fast in practice, consistently outperforming Hopcroft's algorithm (HOP). This result is surprising, given the simplicity of the algorithm. The implementation of this algorithm constructs an intermediate DFA; the performance can be further improved by eliminating this intermediate step.

- The new algorithm (BW) displayed excellent performance. The algorithm is frequently faster than even Brzozowski's algorithm. Unfortunately, this algorithm can be erratic at times — not surprising given its exponential running time. The algorithm can be further improved using memoization.

Given these conclusions, we can make the following recommendations:

1. Use the new algorithm (BW, appearing as [Wat95, Algorithm 7.28]), especially in real-time applications (see [Wat95, Chapter 7] for an explanation of why this algorithm is useful for real-time applications). If the performance is still insufficient, modify the algorithm to make greater use of memoization.

2. Use Brzozowski's algorithm (derived in [Brzo62] and [Wat95, Chapter 7]), especially when simplicity of implementation or consistent performance is desired. The algorithm is able to deal with an FA as input (instead of only DFAs), producing the minimal equivalent DFA. When a minimization algorithm is being combined with a FA construction algorithm, Brzozowski's minimization algorithm is usually the best choice. The DFA construction algorithms are usually significantly slower than the FA construction algorithms, as was shown in the previous sections. For this reason, a FA construction algorithm combined with Brzozowski's minimization algorithm will produce the minimal DFA faster than a DFA construction algorithm combined with any of the other minimization algorithms.

   Brzozowski's algorithm can be further improved by eliminating the DFA which is constructed in an intermediate step.

3. Use Hopcroft's algorithm for massive DFAs. It is not clear from the data in this paper precisely when this algorithm becomes more attractive than the new one or Brzozowski's.

4. The two most-commonly taught text-book algorithms (the Aho-Sethi-Ullman algorithm and the Hopcroft-Ullman algorithm) do not appear to be good choices for high performance. Even for simplicity of implementation, Brzozowski's algorithm is better.

5 Future directions for the toolkit

A number of improvements to FIRE Lite will appear in future versions:

- Presently, FIRE Lite implements only acceptors. Transducers, such as Moore and Mealy machines (as would be required for some types of pattern matching, lexical analysis, and communicating finite automata), will be implemented in FIRE Lite and will be described at the workshop.
- A future version of the toolkit will include support for extended regular expressions, i.e. regular expressions containing intersection or complementation operators, and for linear (regular) grammars.
- Basic regular expressions and automata transition labels are represented by character ranges. A future version of FIRE Lite will permit basic regular expressions and transition labels to be built from more complex data-structures. For example, it will be possible to process a string (vector) of structures.

The future directions for the toolkit are presented in more detail in [Wat96] and at the Ribbit Software Systems Inc. web site, www.RibbitSoft.com.

6 FIRE Lite experiences and conclusions

A large number of people (worldwide) have made use of the FIRE Engine, and a number of people have started using FIRE Lite. As a result, a great deal of experience and feedback has been gained with the use of the finite automata toolkits. Some of these are listed here.

- The FIRE Engine and FIRE Lite toolkits were both created before the SPARE Parts. Without experience writing class libraries, it was difficult to devise a general purpose toolkit without having a good idea of what potential users would use the toolkit for. FIRE Lite has evolved to a form which now resembles the SPARE Parts (for example, the use of call-back functions).
- The FIRE Engine interface proved to be general enough to find use in the following areas: compiler construction, hardware modeling, and computational biology. The additional flexibility introduced with the FIRE Lite (the call-back interface and multi-threading) promises to make FIRE Lite even more widely applicable.
- Thanks to the documentation and structure of the FIRE Engine and FIRE Lite, they have both been useful study materials for students of introductory automata courses.
- The SPARE Parts was developed some two years after the taxonomy in [Wat95, Chapter 4] had been completed. By contrast, FIRE Lite was constructed concurrently with the taxonomy presented in [Wat95, Chapter 6]. As a result, the design phase was considerably more difficult (than for the SPARE Parts) without a solid and complete foundation of abstract algorithms.
- After maintaining and modifying several upgrades of the FIRE Engine, FIRE Lite is likely to be considerably easier to maintain and enhance.

Acknowledgements: I would like to thank Darrell Raymond, Derick Wood, Richard Watson, Nanette Saes, and Caroline Sori for their assistance in preparing this paper.

References

[ASU86] AHO, A.V., R. SETHI, and J.D. ULLMAN. *Compilers: Principles, Techniques, and Tools*. (Addison-Wesley, Reading, MA, 1988).

[Brzo62] BRZOZOWSKI, J.A. Canonical regular expressions and minimal state graphs for definite events, in: *Mathematical theory of Automata, Vol. 12 of MRI Symposia Series*. (Polytechnic Press, Polytechnic Institute of Brooklyn, NY, 1962) 529–561.

[BS86] BERRY, G. and R. SETHI. From regular expressions to deterministic automata, *Theoretical Computer Science* 48 (1986) 117–126.

[CH91] CHAMPARNAUD, J.M. and G. HANSEL. Automate: A computing package for automata and finite semigroups, *J. Symbolic Computation* 12 (1991) 197–220.

[Hopc71] HOPCROFT, J.E. An $n \log n$ algorithm for minimizing the states in a finite automaton, in: Z. Kohavi, ed., *The Theory of Machines and Computations*. (Academic Press, New York, 1971) 189–196.

[HU79] HOPCROFT, J.E. and J.D. ULLMAN. *Introduction to Automata, Theory, Languages, and Computation*. (Addison-Wesley, Reading, MA, 1979).

[John86] JOHNSON, J.H. INR: A program for computing finite automata, Technical Report, Department of Computer Science, University of Waterloo, Waterloo, Canada, January 1986.

[JPTW90] JANSEN, V., A. POTHOFF, W. THOMAS, and U. WERMUTH. A short guide to the Amore system, *Aachener Informatik-Berichte* 90(02), Lehrstuhl für Informatik II, (Universität Aachen, January 1990).

[Leis77] LEISS, E. Regpack: An interactive package for regular languages and finite automata, Research Report CS-77-32, Department of Computer Science, University of Waterloo, Waterloo, Canada, October 1977.

[MY60] MCNAUGHTON, R. and H. YAMADA. Regular expressions and state graphs for automata, *IEEE Trans. on Electronic Computers* 9(1) (1960) 39–47.

[PTVF92] PRESS, W.H., S.A. TEUKOLSKY, W.T. VETTERLING and B.P. FLANNERY. *Numerical Recipes in C: The Art of Scientific Computing*. (Cambridge University Press, Cambridge, England, 2nd edition, 1992).

[RW93] RAYMOND, D.R. and D. WOOD. The Grail papers, Department of Computer Science, University of Waterloo, Canada. Available by ftp from CS-archive.uwaterloo.ca.

[Thom68] THOMPSON, K. Regular expression search algorithms, *Comm. ACM* 11(6) (1968) 419–422.

[Wat93a] WATSON, B.W. A taxonomy of finite automata construction algorithms, Computing Science Report 93/43, Eindhoven University of Technology, The Netherlands, 1993. Available for ftp from ftp.win.tue.nl in directory /pub/techreports/pi/automata/.

[Wat93b] WATSON, B.W. A taxonomy of finite automata minimization algorithms, Computing Science Report 93/44, Eindhoven University of Technology, The Netherlands, 1993. Available for ftp from ftp.win.tue.nl in directory /pub/techreports/pi/automata/.

[Wat94a]   WATSON, B.W. An introduction to the FIRE Engine: A C++ toolkit for FInite automata and Regular Expressions, Computing Science Report 94/21, Eindhoven University of Technology, The Netherlands, 1994. Available for ftp from `ftp.win.tue.nl` in directory `/pub/techreports/pi/automata/`.

[Wat94b]   WATSON, B.W. The design and implementation of the FIRE Engine: A C++ toolkit for FInite automata and Regular Expressions, Computing Science Report 94/22, Eindhoven University of Technology, The Netherlands, 1994. Available for ftp from `ftp.win.tue.nl` in directory `/pub/techreports/pi/automata/`.

[Wat95]   WATSON, B.W. *Taxonomies and Toolkits of Regular Language Algorithms*. (Eindhoven University of Technology, The Netherlands, ISBN 90-386-0396-7, 1995).

[Wat96]   WATSON, B.W. Implementing and using finite automata toolkits, in: Wahlster, W., *Proceedings of the 12th European Conference on Artificial Intelligence* (Budapest, Hungary, August 1996).

Author Index

Lecture Notes in Computer Science

For information about Vols. 1–1186

please contact your bookseller or Springer-Verlag

Vol. 1224: M. van Someren, G. Widmer (Eds.), Machine Learning: ECML-97. Proceedings, 1997. XI, 361 pages. 1997. (Subseries LNAI).

Vol. 1225: B. Hertzberger, P. Sloot (Eds.), High-Performance Computing and Networking. Proceedings, 1997. XXI, 1066 pages. 1997.

Vol. 1226: B. Reusch (Ed.), Computational Intelligence. Proceedings, 1997. XIII, 609 pages. 1997.

Vol. 1227: D. Galmiche (Ed.), Automated Reasoning with Analytic Tableaux and Related Methods. Proceedings, 1997. XI, 373 pages. 1997. (Subseries LNAI).

Vol. 1228: S.-H. Nienhuys-Cheng, R. de Wolf, Foundations of Inductive Logic Programming. XVII, 404 pages. 1997. (Subseries LNAI).

Vol. 1230: J. Duncan, G. Gindi (Eds.), Information Processing in Medical Imaging. Proceedings, 1997. XVI, 557 pages. 1997.

Vol. 1231: M. Bertran, T. Rus (Eds.), Transformation-Based Reactive Systems Development. Proceedings, 1997. XI, 431 pages. 1997.

Vol. 1232: H. Comon (Ed.), Rewriting Techniques and Applications. Proceedings, 1997. XI, 339 pages. 1997.

Vol. 1233: W. Fumy (Ed.), Advances in Cryptology — EUROCRYPT '97. Proceedings, 1997. XI, 509 pages. 1997.

Vol 1234: S. Adian, A. Nerode (Eds.), Logical Foundations of Computer Science. Proceedings, 1997. IX, 431 pages. 1997.

Vol. 1235: R. Conradi (Ed.), Software Configuration Management. Proceedings, 1997. VIII, 234 pages. 1997.

Vol. 1236: E. Maier, M. Mast, S. LuperFoy (Eds.), Dialogue Processing in Spoken Language Systems. Proceedings, 1996. VIII, 220 pages. 1997. (Subseries LNAI).

Vol. 1238: A. Mullery, M. Besson, M. Campolargo, R. Gobbi, R. Reed (Eds.), Intelligence in Services and Networks: Technology for Cooperative Competition. Proceedings, 1997. XII, 480 pages. 1997.

Vol. 1239: D. Sehr, U. Banerjee, D. Gelernter, A. Nicolau, D. Padua (Eds.), Languages and Compilers for Parallel Computing. Proceedings, 1996. XIII, 612 pages. 1997.

Vol. 1240: J. Mira, R. Moreno-Díaz, J. Cabestany (Eds.), Biological and Artificial Computation: From Neuroscience to Technology. Proceedings, 1997. XXI, 1401 pages. 1997.

Vol. 1241: M. Akşit, S. Matsuoka (Eds.), ECOOP'97 – Object-Oriented Programming. Proceedings, 1997. XI, 531 pages. 1997.

Vol. 1242: S. Fdida, M. Morganti (Eds.), Multimedia Applications, Services and Techniques – ECMAST '97. Proceedings, 1997. XIV, 772 pages. 1997.

Vol. 1243: A. Mazurkiewicz, J. Winkowski (Eds.), CONCUR'97: Concurrency Theory. Proceedings, 1997. VIII, 421 pages. 1997.

Vol. 1244: D. M. Gabbay, R. Kruse, A. Nonnengart, H.J. Ohlbach (Eds.), Qualitative and Quantitative Practical Reasoning. Proceedings, 1997. X, 621 pages. 1997. (Subseries LNAI).

Vol. 1245: M. Calzarossa, R. Marie, B. Plateau, G. Rubino (Eds.), Computer Performance Evaluation. Proceedings, 1997. VIII, 231 pages. 1997.

Vol. 1246: S. Tucker Taft, R. A. Duff (Eds.), Ada 95 Reference Manual. XXII, 526 pages. 1997.

Vol. 1247: J. Barnes (Ed.), Ada 95 Rationale. XVI, 458 pages. 1997.

Vol. 1248: P. Azéma, G. Balbo (Eds.), Application and Theory of Petri Nets 1997. Proceedings, 1997. VIII, 467 pages. 1997.

Vol. 1249: W. McCune (Ed.), Automated Deduction – CADE-14. Proceedings, 1997. XIV, 462 pages. 1997. (Subseries LNAI).

Vol. 1250: A. Olivé, J.A. Pastor (Eds.), Advanced Information Systems Engineering. Proceedings, 1997. XI, 451 pages. 1997.

Vol. 1251: K. Hardy, J. Briggs (Eds.), Reliable Software Technologies – Ada-Europe '97. Proceedings, 1997. VIII, 293 pages. 1997.

Vol. 1252: B. ter Haar Romeny, L. Florack, J. Koenderink, M. Viergever (Eds.), Scale-Space Theory in Computer Vision. Proceedings, 1997. IX, 365 pages. 1997.

Vol. 1253: G. Bilardi, A. Ferreira, R. Lüling, J. Rolim (Eds.), Solving Irregularly Structured Problems in Parallel. Proceedings, 1997. X, 287 pages. 1997.

Vol. 1254: O. Grumberg (Ed.), Computer Aided Verification. Proceedings, 1997. XI, 486 pages. 1997.

Vol. 1255: T. Mora, H. Mattson (Eds.), Applied Algebra, Algebraic Algorithms and Error-Correcting Codes. Proceedings, 1997. X, 353 pages. 1997.

Vol. 1256: P. Degano, R. Gorrieri, A. Marchetti-Spaccamela (Eds.), Automata, Languages and Programming. Proceedings, 1997. XVI, 862 pages. 1997.

Vol. 1258: D. van Dalen, M. Bezem (Eds.), Computer Science Logic. Proceedings, 1996. VIII, 473 pages. 1997.

Vol. 1259: T. Higuchi, M. Iwata, W. Liu (Eds.), Evolvable Systems: From Biology to Hardware. Proceedings, 1996. XI, 484 pages. 1997.

Vol. 1260: D. Raymond, D. Wood, S. Yu (Eds.), Automata Implementation. Proceedings, 1996. VIII, 189 pages. 1997.

Vol. 1262: M. Scholl, A. Voisard (Eds.), Advances in Spatial Databases. Proceedings, 1997. XI, 379 pages. 1997.

Vol. 1263: J. Komorowski, J. Zytkow (Eds.), Principles of Data Mining and Knowledge Discovery. Proceedings, 1997. IX, 397 pages. 1997. (Subseries LNAI).

Vol. 1264: A. Apostolico, J. Hein (Eds.), Combinatorial Pattern Matching. Proceedings, 1997. VIII, 277 pages. 1997.

Vol. 1266: D. Leake, E. Plaza (Eds.), Case-Based Reasoning Research and Development. Proceedings, 1997. XIII, 648 pages. 1997 (Subseries LNAI).

Vol. 1268: W. Kluge (Ed.), Implementation of Functional Languages. Proceedings, 1996. XI, 284 pages. 1997.

Vol. 1270: V. Varadharajan, J. Pieprzyk, Y. Mu (Eds.), Information Security and Privacy. Proceedings, 1997. XI, 337 pages. 1997.